Managing CHANGE for Safety & Health Professionals

A 6 Step Process

F. David Pierce, CSP, CIH

Government Institutes
Rockville, MD

Government Institutes, Inc., 4 Research Place, Rockville, Maryland 20850, USA.

01 00 99 98 97 5 4 3 2 1

Library of Congress Cataloging-in-Publication Data

Pierce, F. David.
Managing change for safety and health professionals: a six step process for achieving total quality / F. David Pierce.
p. cm.
Includes index.
ISBN: 0-86587-563-4
1. Industrial safety-Management. 2. Total quality management.
I. Title.
T55.P52 1997
658.3' 82--dc21

96-37927
CIP

Printed in the United States of America

MANAGING CHANGE FOR SAFETY AND HEALTH PROFESSIONALS

CONTENTS

Preface

Change isn't just a factor in our lives. Dealing with it has become the secret to our survival. Change has become exponential. During the first half of the 20th century, major changes came about gradually: automobiles, airplanes, radios, television, nuclear energy, electricity, and the first computers.

The second half of the century has brought changes that seem to happen daily and at an ever-increasing magnitude: the transistor, the microchip, fax machines, modems, VCR's, CD's, microwave ovens, cellular telephones, notebook computers, satellite paging, cable and satellite television, bungie jumping, suburban malls, interstate highways, smart missiles, lasers, CAT scans, arthroscopic surgery, OSHA, EPA, EEO, ADA, Affirmative Action, Martin Luther King Jr., microwave transmission, video rentals, fast food, telecommuting, smog, transplant surgery, the ozone hole, the Greenhouse Effect, personal computers, acid rain, waste to energy conversion, flat organizations, outsourcing, team dynamics, ergonomics, corporate coaching, workplace violence, TQM, TPM, TBM, JIT, W. Edwards Demming, the Malcolm Baldrige Award, ISO 9000 and 14000, re-engineering, word processing, database management, networks, terrorism, pizza delivery, the quartz watch, ESPN, CNN, copy machines, one hour dry cleaning, gene splicing, the internet, photo IDs, cumulative trauma disorders, teleconferencing, stress syndromes, micro-breweries, restaurant chains, mutual funds, no-load funds, NASDAQ, mass mailing, the Peoples' Choice Awards, MTV, rock and roll, rap, electronic synthesizers, CAD, computer graphics, telemarketing. The list goes on and on. These changes don't represent new ideas as much as they represent revolutions in thought, practicc, and expectation.

With all this change around us, we've surely become used to and at ease with change. Then, why is it that change is so disruptive? Why is it that when it's our turn to implement a major change it so often goes poorly? Why is it that the most common reactions to change are ignoring or opposing it? If change is so critical for our ability to keep up and make profits in the future, why is change so difficult for us?

When we try to understand change, we must first begin to understand how we respond to it individually and why. Although normal, these reactions are not genetically encoded. Next, we need to understand how we reject, accept, and explore change. This appears to be a hard-wired pattern. We need to understand our role options in change. As with most things in life, we have choices. True, some choices are easier to identify as being better than others, but they are choices we make. And because change is certainly not going to go away, we need to understand how to successfully implement change. We interface within our organizations daily. We are in positions to make positive changes in the way our functions perform and how they work in the organization. Most often this is major change. If we are to be successful in today's and especially tomorrow's world, we must become skilled change masters.

After twenty-five-plus years of implementing change and helping companies do the same, I've come to an inescapable conclusion. Successfully implementing change is a skill. More than that, it is a skill that few leaders in our organizations possess. Consequently, major change is most often done poorly. No wonder we have a natural skepticism and resistance to change. But the word still bears hope for us. Like computer use or the practice of our specialties, change management can be learned.

I wrote this book to clear the confusion about change and to offer a successful method for implementing change. After watching thousands react to change and studying why hundreds of change efforts have been successful (and just as importantly, why many fail), I have devised a method for successful change implementation. It will work as effectively if you are inserting change into business, government, academia, or a Girl Scout troop.

Change needn't be overly disruptive. It certainly doesn't need to be frightening. Change is critical to tomorrow's success. Skills for effectively implementing change will be essential for those who will lead us into tomorrow. You could be such a leader. Are you ready for the challenge?

F. David Pierce
Sandy, Utah

About the Author

F. David Pierce is a Certified Safety Professional and Certified Industrial Hygienist with over 25 years experience in industry and government. He is president of Leadership Solution Consultants, Inc., of Salt Lake City, Utah, a change management and safety excellence consulting firm.

Pierce's experience includes 12 years with Westinghouse Electric Corporation's Western Zirconium Plant, where, as manager of safety, environment and security, he spearheaded the efforts which resulted in a Malcolm Baldridge National Quality Award.

Pierce is the author of *Total Quality for Safety and Health Professionals* (Government Institutes, 1995), and *Shifting Safety and Health Paradigms* (Government Institutes, 1996). He is a frequent speaker at professional conferences and teaches popular courses on Total Quality and Occupational Safety and Health Management for the National Institute for Occupational Safety and Health's Education Resource Centers.

Pierce received his Master of Science degree in public health from the University of Utah–Salt Lake City. He is past president of the Utah chapter of the American Society of Safety Engineers and former chair of the Utah Occupational Safety and Health Advisory Council. Pierce currently chairs the Utah Safety and Hygiene Conference Committee and has served on the advisory committee for the Rocky Mountain Center for Occupational Safety and Health. He is a member of the American Industrial Hygiene Association, American Society of Safety Engineers, American Academy of Industrial Hygiene, International Hazardous Materials Association, and National Association of Environmental Professionals.

Dedication

To Ruth and Neal, my "chosen" parents who by intuition saw something positive and promising in a skinny kid years ago, took him into their family and treated him as one. Through your success in dealing with seemingly constant change, you brought stability where crisis threatened and commitment where overwhelming challenge awaited.

Introduction: Change

When I was young, I helped my grandfather on his dairy farm each summer. It didn't seem like much of a vacation at the time. Being young, there were limits to what I could do, but I worked hard—starting long before sunup and continuing after sundown. Exhausted at the end of the day, I would simply fall into bed. Sleep was not a problem. Thinking back on this quality time with my grandparents, oddly enough, what I remember most isn't the hard work, exhaustion, and long hours. And it isn't the constant interruptions caused by breakdowns and quick fixes that operating an aging dairy farm involved. Instead, what I remember most are the quiet times, sitting on a hay bale or a wagon eating lunch or just listening to my grandfather. I remember sitting in my grandfather's small living room at the foot of his rocking chair responding to his questions and being graced with his gifts of wisdom about life.

My grandfather was a well-educated man. He loved the farm but was also an ordained minister and a teacher at the local college. He was a philosopher and, in his own way, a visionary of the future. He recollected for me the many changes that had occurred during his lifetime (1877–1960). He had built the house my grandparents lived in—and the house before that—with his own hands. In his lifetime, electricity had replaced candles and lanterns. Plumbing changed from a pump and an outhouse to the convenience of indoor running water. Heating a home became as easy as turning up the thermostat, although he still preferred the fireplace and a goose down quilt. You could go to the store and buy lumber instead of going into the nearby forest and cutting it yourself. You no longer had to grow most of the food that you and your family needed. Music could be brought home and played on a phonograph. Voices could be magically transported through the air and come out of a radio. The voices became images through television. Cars and trucks replaced the horse and wagon and by these vehicles, the miles between people and places melted away. Airplanes transported people around the world. Telephones came to every home and made travel nonessential to communication. Three officially declared wars had come and gone. Because of his commitments, he had not served in the military, but friends had and some hadn't returned. He had experienced the advancement of warfare from the crude explosion and sinking of the USS Maine in Havana harbor to the cataclysmic and instant technological vaporization of two cities in Japan. The changes that he saw and adapted to are rather impressive.

Even with all the changes that happened in my grandfather's lifetime, however, there was a quiet peacefulness about the times in which he lived. People just weren't in a huge hurry. As changes came about, people generally took them in stride. The expected, "Oh, wow!" reaction to these changes was a more subdued; "How about that?" There was a patience about change.

Today, we are anything but patient about change. The rat race we are all caught up in today really didn't start until the 1960s or the early 1970s. Our pace became faster and continues to speed up. Our patience is growing even scarcer. Changes come at us almost daily and are as large in magnitude as those in my grandfather's lifetime. We have become so used to change and desensitized to quantum leaps in change, that we at times hardly notice it. Change is expected.

CHANGE IN TODAY'S WORLD

In fact, the world is changing as never before. These changes have direct impacts on us personally, on our jobs, and on every aspect of our lives. How much change is occurring? Let's look at some of the most profound recent changes.

Changes in Where We Work. According to Price Pritchett in *New Work Habits*, in the early 1900s, about 85% of America's workers were employed in agriculture. Today, agriculture employs less than 3% of our workforce. One-half of all workers in the world's industrialized nations were involved in manufacturing in the 1960s. Today, no industrial country has more than about one-sixth of its workers in manufacturing. It is estimated that about two-thirds of America's workers are employed in services today. The Department of Labor estimates that by the year 2000, at least 44% of the workforce will be in the information business. Jobs change. Today, however, job change occurs much faster and in areas of our economy that didn't even exist 20 years ago.

Changes in How We Work. Charles Handy tells us in *The Age of Reason* that less than half of the workers in the world's industrial nations will be holding full-time jobs by the beginning of the 21st century. Every year, more and more of us are becoming self-employed. Additionally, as this continues to change, many will choose temporary or part-time employment because it meets their needs or because it is all that is available. According to Mary O'Hara-Devereaux and Robert Johansen in *Global Work*, in 1991, one out of three American workers had been with their employer less than a year and two out of three for less than five years. Our contingent workforce now equals about 45,000,000 consultants and temporary, part-time, or self-employed workers. That number has grown 57% since 1980. The 9-to-5 workday is all but obsolete, as are lifetime jobs, predictability, hierarchical relationships, corporate security blankets, and for a growing sector of our workers, the workplace itself. For tomorrow's workforce, job-hopping will be the norm.

Changes in Our Businesses. Businesses that are successful are constantly changing. If you look at a Fortune 100 listing of American companies from the early 1900s, less than 16 are still in business. From *Fortune* magazine's first list of America's biggest 500 companies in 1956, only 29 of those top 100 businesses are still in the top 100 today. By 1990, 230 of the companies (46%) on the 1980 Fortune 500 disappeared. In today's business world, size or past reputation guarantees nothing. In addition, business markets are being created in areas where no business existed just 10 years ago. There has never been a greater need for companies to expand into new markets or make quantum leaps in product advancement. Those that haven't advanced never receive a second chance because of the speed at which business changes.

Changes in Technology. Ian Morrison and Greg Schmid in *Future Tense* state that the integrated circuit invented in the late 1950s was responsible for beginning the Information Revolution. There has been an ever-increasing movement to store more and more on a single microchip. Between 1960 and 1970, the number of components on a single microchip doubled each year from 1 to 1000.

Since 1970, the number of components have doubled every year and a half reaching 1,000,000 in 1990 and 1,000,000,000 in 1992. They state that today's average consumer wears more computing power on his or her wrist than existed in the entire world before 1961. In addition, the first practical robot was introduced into business in the 1960s. By 1982, there were an estimated 32,000 robots used in American business. Today there are over 20,000,000.

Everything is smaller, faster and cheaper. ENIAC, which in 1944 was commonly thought to be the first modern computer, took up more space than an 18-wheeler's trailer, weighed more than 17 Chevrolet Camaros, and consumed 140,000 watts of electricity, according to *Business Week* in "The Information Revolution." All this so that ENIAC could perform up to 5,000 basic arithmetic computations per second. Today's 486 microprocessor is constructed from a tiny piece of silicon about the size of a dime, weighs less than a packet of Sweet 'N Low, uses less than 2 watts of power, and can execute up to 54,000,000 instructions per second. The cost of computers drops about 30% every year and the microchips are doubling in performance every 18 months.

Computer power is 8,000 times less expensive than it was 30 years ago. John Naisbitt recorded in *Global Paradox* an analogy used by Randall Tobias, former Vice Chairman of AT&T: "If we had similar progress in automotive technology, today you could buy a Lexus for about $2. It would travel at the speed of sound, and go about 600 miles on a thimble of gas."

Peter Large tells us in *The Micro Revolution Revisited* that, "You're in Paris, and you decide to use your American Express card. Getting credit approval involves a 46,000-mile journey over phone lines and computers. The job can be completed in 5 seconds."

John Huey used other analogies to emphasize this point in *Fortune* magazine. "Let's say you're going to a party, so you pull out some pocket change and buy this little greeting card that plays 'Happy Birthday' when opened. After the party, someone casually tosses the card into the trash, throwing away more computer power than existed in the entire world in 1950. The home video camera you use to take pictures of the party contains more processing power than an old IBM 360, the wonder machine that gave birth to the mainframe computer age. The party gift you give is a system called Saturn, made by Sega, the gamemaker. It runs on a higher-performance processor than the original 1976 Cray supercomputer, which in its day was accessible to only the most elite physicists."

Our use and dependence on technology is changing. Since as recently as 1983, American business has added 25,000,000 computers. Cellular telephone users went from zero in 1983 to 16,000,000 by the end of 1993. 19,000,000 of us now carry pagers and about 12,000,000,000 messages were left in voice mailboxes in 1993 alone. Since 1987, homes and businesses purchased 10,000,000 fax machines and E-mail addresses increased by over 26,000,000, all according to Rich Tetzeli in *Fortune*.

Changes in Information. Our dependence on information is changing. Communication technology is changing so radically that the speed and direction of information flow is altering our work roles. For example, the number of secretaries has decreased by over half a million just since 1987.

"There has been more information produced in the last 30 years than during the previous 5,000. A weekday edition of *The New York Times* contains more information than the average person was likely to come across in a lifetime during 17th-century England. The information supply available to us doubles every 5 years" (Richard Saul Wurman in *Information Anxiety*.)

In 1991, for the first time ever, businesses spent more on computing and communications equipment than the combined amount spent on industrial, mining, farm, and construction equipment. According to Price Pritchett, this spending pattern offers proof that we have entered a new era. The Information Age has taken the reins from the Industrial Age.

HOW ORGANIZATIONS VIEW THEIR FUTURES

Organizations are caught up in this change also. But this whirlwind of change not only impacts the actual organizations, it impacts how they view their future. In the early 1970s, America's businesses saw the role of change in their future as follows:

No change 60%
Sporadic, incremental change 35%
Continuous, overlapping change ... 5%

Arguably, those who saw the future to include continuous, overlapping change were visionary and those who saw no need to change in order to ensure continued success were myopic. This belief, however, changed. Those who were falsely complacent about the necessity of change, have themselves changed. This same poll today has vastly different results.

No change 1%
Sporadic, incremental change 24%
Continuous, overlapping change .. 75%

One has to ask, are businesses ready for such a constantly changing environment? Daryl Connor, in *Managing at the Speed of Change*, referred to this changing environment as the "Houseboat Fallacy." He likened our traditional management to that of piloting a houseboat on a calm lake with many safe side canyons for seeking refuge should a storm come. He likened today's constantly changing business environment to piloting that same houseboat down the rapid-filled white water of a turbulent flowing river. "Hang on," management assures the crew. Things will slow down soon when we hit the calm waters of the lake. In reality, though, the calm never comes. Riding the new environment of the river is the new, inescapable reality.

Connor's point is that success today requires new skills and new organizations. Obviously, a houseboat is at a considerable disadvantage on a fast-moving river. The craft (organization) needs to be nimble (efficient), quick in response (communicative and highly participative), have a shallow draft (versatile changing culture), and be difficult to capsize (resilient and motivated). American business is attempting to redesign the "craft." The problem is that a standardized set of drawings or redesigned methods do not exist. Without them, businesses resort to traditional methods and a great diversity of outcomes result.

But, Connor said, the change in the organization is only half of the key to future success. New skills are also needed. This is not only a challenge to top management; it is a challenge to everyone who is to be successful piloting or crewing the newly designed boat. It is a challenge to all of us.

OUR ROLE IN CHANGE

The signposts are all around us: Everything is changing. Like it or not, we are caught up in this rapidly changing environment. Our organizations are changing, as are businesses in general. Why then do we cling to our traditional ways of performing our safety and health tasks? If everything else is changing, how can we expect our programs, turf, or approach to remain untouched by change? Whether we cling to keeping the "teeth" in the regulatory agencies or to fighting downsizing of safety and health staffing, these are merely signposts along this journey we call change. In a changing environment, nothing can remain untouched or unshaken. Because of this, change is often frightening. But, just as you shake a kaleidoscope to change the pattern, lifting an ever-changing world to our eyes continually amazes us with its new patterns. Change and the new patterns it brings can be beautiful, challenging, and exciting to those who develop an ease about change. These people, of course, will be the survivors and leaders in this emerging world.

Obviously, there are two roles we can play in an ever-changing environment. Most of the time, we make a conscious decision about which role to take. We can choose to be acted on by change. Choosing to be mere leaves in the winds of change, however, means we settle where the changing forces place or bury us. There is an alternative. We can choose to be leaders in change, a course that brings more uncertainties and challenges, but offers us some, if not a great deal of, influence on where we end up. Uncertainties include what mechanism of change we use, what direction we pursue, and how we plan and implement change. But who best to orchestrate the change in our specialties than ourselves? Who else could possibly have the potential vision with an unyielding grip on our mission for worker safety and health? In this aspect of our changing world, we must be leaders.

A colleague of mine recently asked me if I knew of any job openings. "Thinking of leaving yours?" I asked. His reply was quiet and simple. "No, I'm no longer employed."

I asked, "What happened, John?" "I guess the company didn't think safety and health was that important anymore," he responded. From our continuing conversation I learned that the company he had worked for was going through some fundamental changes. To use the method's current name, they were "reengineering" the organization. Middle management had been cut. Span of control for supervisors and managers had been expanded. Within this process, the number of workers in nonproduction areas was being reduced as well. Fundamental questions were asked by management, such as "Can we afford a particular function or a certain level of a particular function?" and "What level of service or function do we need and are there more economical or more effective ways of providing it?" To my safety and health colleague, these questions sounded like threats.

"What actions did you take to help answer the questions?" I asked. His approach had been very traditional. Our profession had taught him well. He was trained to be reactive when questions were asked of him directly. When he was asked he talked about the costs of an accident or a citation from OSHA. Obviously, his approach did not effectively communicate any need above the economic downsizing demands of the company. After all, it is hard to sell the need for vaccination when the patient is dying from blood loss.

My colleague suffered from a common ailment of our profession—circling the wagons. He could not see beyond his traditional paradigms of how a safety and health program has always been done. His reward for being traditional was an "outplacement opportunity." It's a shame, too. Any company going through a significant change desperately needs someone on the inside leading and guiding the process so that the importance of worker safety and health is not lost or devalued.

Our role in change is critical. It is critical from a worker safety and health perspective, and just as my safety and health colleague discovered, on a personal level as well.

Recently, the *Wall Street Journal* published articles that have been very critical of the approaches American businesses are taking in their reinvention efforts and how deeply they are cutting costs. In fact, a dedicated focus on streamlining American business and reducing costs is necessary for our country's business to survive in the global economy. This point is not in question. The methods that American businesses take are both traditional and individual in their selection. But, we are a zealous people. Seldom do we have the wisdom to know when "enough is enough." We have a long history of going too far, too fast. The current term to describe this state is "corporate anorexia."

This state, the methods each business chooses for change, or the value of that process, is not within the scope of this book. This book, instead, is about two things that are critical to our practice and to worker safety and health in America—our survival and gaining efficiency in our practice. The short-term plight of worker safety and health in this country will be adversely affected if our profession becomes extinct. There is too much collective knowledge to begin again. This is why our chosen role in change is paramount to worker safety and health in America. We literally stand at the crossroads of what will determine the viability of our profession and whether we will gain or lose ground in worker safety and health. Knowing this, I can think of no better time to talk about change, our role in that change, and our transition to efficiency.

Managing CHANGE *for* Safety & Health Professionals

A 6 Step Process

1

Moving Toward Efficiency

American business is changing. The need to change with it is undeniable. But the direction and method of change are subject to debate. With no agreed-upon common vision, American business seems to be heading in many directions at once. This isn't surprising. The direction (intermediate vision) each business takes is dependent upon the culture of that business, its markets, management paradigms, and practices, and what are traditional beliefs to Americans. With this diversity in direction, it is obvious that some businesses will change effectively to meet the challenges of tomorrow's business world and others will not. The survival of a business has always been dependent on this fact.

There are some common and easily discernible similarities in all businesses undergoing change. First, change begins with cost-focused cutting and reorganization. The American method of choice has been downsizing, or what is currently called reengineering. This is an oversimplification, but one that is appropriate for our use. Second, several common changes are being fostered across the board in these changing businesses. This is because these common changes are working and the unanimous opinion is that the old ways don't work. These common changes include a higher level of worker participation in the business process, the use of team dynamics, better planning and measurement of performance, elimination of waste and wasted efforts, increased line responsibilities and span of control, the use of benchmarking and partnershipping, and a new focus on continual improvement versus the traditional technological quantum leap.

Unless you've been stranded on a desert island for the past few years, you probably recognize these common changes as being elements of what we have come to know as total quality management (TQM). But, now labeled, we have some predictable paradigms to transcend. "Total quality was a fad we went through a few years ago, wasn't it?" and "TQM was merely a buzz word that really didn't work" are common thoughts. Now that we are past the fad-chasing phase of TQM, let's set the record straight. In over 95% of cases, the management fad attempted under the name TQM was not total quality at all. Most often, it was an Americanized creation that agreed with our traditional management concepts and used some of or most of the TQM buzz words. This is the number-one reason why most U.S. total quality initiatives failed. There are three other significant reasons why total quality attempts have failed. 1) Management only viewed TQM as a fad, so they didn't really get serious about it. A halfhearted effort always yields nothing. 2) Total quality was too contrary to traditional American management practice to be accepted by management. Traditional ways were so entrenched that total quality was quickly rejected. And, 3) the true concepts of TQM were Americanized to the point that they no longer resembled total quality. In these cases, what was implemented either didn't work or was so close to the way business was already being done that no real change occurred.

"In today's changing world, it's important to be known as being supportive of the organization's efforts. That's why you get 'management points' when you use their buzz words."

"I told my boss, 'I'd like to benchmark their value system for insertion into my continuous improvement plan.' He loved it."

In some companies, total quality was successfully implemented and brought about a recognizable change in those organizations. The success of those total quality organizations conclusively proves the effectiveness of total quality. In fact, most companies that implemented total quality experienced quantum advancement in their business sectors. But, like businesses in general, some total quality companies failed. That percentage, however, is much smaller than the failure rate for those that held on to traditional management styles.

Total quality is not the same as corporate downsizing; it is a dynamic management concept that continually evolves by its own concepts and components. This fact alone provided a considerable paradigm shift for American businesses. Before looking at these undeniable common changes in American businesses as total quality concepts, we should stop and take a new look at what was known as TQM and our experiences with it.

STUMBLING BLOCKS

One of the major difficulties in understanding TQM was the concept of quality: the difference between "Big Q" and "little q." When Americans think of quality, we think of Big Q. Big Q is quality inspections, quality assurance, statistical process control, quality specifications, quality procedures, etc. These are the basis of the Malcolm Baldridge Award, a national quality prize. Little q is very different. It is *the way things are done*, the approach, the processes that produce the quality product or service that customers expect.

The difference is important, but it can be confusing. Consider this example. A new car rolls off the assembly line. Throughout assembly, it has gone through many quality checks, inspections, and tests according to quality procedures. Quality assurance documents detail the results. The car simply cannot leave the plant without this paperwork being complete and in order. This is Big Q. On the other hand, when you go to the local car dealership to look for a new car to buy, you look for little q. Generally, you don't care about the quality procedures, inspections, checks, or test results. If you did, you would insist on seeing them for the car you were considering buying. No one does that. As customers, we just aren't interested in Big Q. We are, however, very interested in little q. We simply want a car that looks terrific, has clean lines and excellent paint, is extremely dependable, needs little maintenance, runs well, has the accessories and performance we want, smells good, and is comfortable to drive. Asked after our purchase if we got a quality car, we would say yes because it fit our definition of little q, not because it passed all the Big Q tests.

Because American business is fixated on Big Q, we have only heard half of W. Edwards Demming's message. We heard all about statistical process control, the need for clear expectations for workers, and the need for clear procedures. But we didn't hear about the need for participation, the different ways we need to manage, that we needed to stop treating our workers as we traditionally have, and about the need for continuous improvement. In fact, through our traditional American business listening filter, we screened out most of what Demming said. We tried to implement less; worse yet, we tried to fit it into the way we have always managed business. No wonder most of our attempts failed.

The second American stumbling block to total quality is in the common abbreviation, TQM. It's common for us to make a term bigger than life until the term takes on a cult-like existence. "Are you a TQM company?" Of course we are and we can use all the other buzz words to talk about how TQM we are. Does the use of TQM phrases have any correlation to actually implementing the concepts and components of total quality? Of course not. We just use the words when they are "in" and to become members of the club. When it's "out," it's *OUT* and we move on to the next fad.

Because of our confusion about Big Q versus little q and our traditional adoration of "in" terms, it may be best to eliminate the term TQM altogether. While we're at it, perhaps we should just eliminate the term "total quality" also. After all, there is nothing magical in the name or the abbreviation. The magic lies in *how* we manage our businesses and programs. We could call it "Demming Management"—although he wasn't the father of total quality and his name is not as influential as it once was since his death. We could call it Japanese Management—although the concept isn't totally Japanese. They certainly have been successful at applying it, however. Calling it Japanese Management would also clash with our made-in-America bias. We could call it

Participative Management—although this may be too restrictive and may tend to limit the term's impact. We could call it Visionary Management—although one can have a vision and still run a business straight into the toilet. We could call it Planning for Improvement—but this title leaves me cold. In reality, we could call it anything except Total Quality and it would in no way diminish what it could bring to a business, organization, or program.

WHY SAFETY AND HEALTH PROFESSIONALS?

When management methods are forced into organizations or programs where there are fundamental mismatches, the attempts at change are totally disruptive and, in many cases, disastrous. These extreme cases are in sharp contrast to those instances where a perfect match exists between the management concept and the organization or program. These positive matches are the opportunities we look for. These are the bright lights we need to run toward, and quickly.

Total quality and safety and health programs are a good match. Total quality management concepts and tools can be woven perfectly into the fabric of a safety and health practice.

OVERVIEW OF TOTAL QUALITY

The Focus of Total Quality. Focus is the correct word: it means an almost religious devotion, a total dedication, a mind-and-body type of effort. Total quality requires a dedicated focus on four things: customers, waste, time, and excellence.

Customers can be external to the organization or program, or they can be internal. These include customers who buy the product or service, those beyond the organization who increase the value of the product or advance production, and those who benefit from that organization or program. From a safety and health perspective, employees and management are examples of inside customers, as are departments that use safety and health services. Think of this focus on customers by drawing a comparison between your safety and health program and a

McDonald's franchise. When customers enter McDonald's, what they see is pleasing—it meets their expectations. Everything is geared to getting customers what they want, the way they want it, in a time frame that meets the customers' needs and the business' needs, for the right cost. Why were drive-up windows conceived? They better met the customers' needs, were more convenient for customers and were generally quicker. They also provided a cost-effective means for meeting the expanding needs of the business. How does this customer focus picture of a McDonald's restaurant compare to your safety and health program?

Total quality focuses on waste. This is not limited to the traditional definition—what goes in the trash can. Waste includes anything that is inefficient or ineffective. Examples include multiple trips made by workers when one will do, too many steps required for people or product, problems with quality, production delays, injuries, illnesses, low productivity, poor or missing procedures, ineffective or missed training, etc. Anything is a waste if it slows down the process or produces a problem with quality. To get a clear picture, think of the inefficiency of getting a new chemical product "approved" in a traditionally run safety and health office. These cumbersome "approval" processes are an excellent example of waste.

Another focus of total quality is time. Time is the only commodity that we have a totally limited supply of. If we need more time, we simply cannot purchase it or borrow it from someone else. There are only 24 hours per day, seven days per week, and 365 days per year. Total quality realizes this and strives to eliminate wasted time. Wasted time includes time waiting for approvals, reports, analyses, someone else, or some step in a process. It includes unnecessary time spent transporting things or ourselves and can include the time we spend doing unnecessary tasks. Time is an important focus of total quality.

Total quality also focuses on excellence. In fact, total quality is a continual effort toward excellence. This is an effort that constantly focuses on a moving target. Total quality knows that as a business or organization gets better, so does the competition. The target, therefore, keeps disappearing over the horizon. Excellence is also difficult to define. It's easier to know what is not excellent than to understand when excellence is achieved. For these reasons, excellence is one of the most difficult things on which to focus. In America, we are accustomed to winning. Winning is the goal in both sports activities and business. Excellence is an unachievable, moving target. This makes winning impossible. Focusing on excellence can only help a business to stay ahead of the competition in that continual effort.

Total Quality and Management. Historically, American management has been authoritative and militaristic. It is built on a hierarchical organizational structure in which management thinks and directs the work and labor does what is directed. Total quality approaches management from a different direction. Total quality is all about leadership. That, of course, is management's job. But, total quality is not autocratic; rather, it is enabling. It models the expected behavior, is encouraging, and doesn't condescend. It is positive, not negative.

Of all the theories of management, total quality is nearest Theory Z. Originally known as Japanese Management or participative management, Theory Z stresses team dynamics. Being led by management, an effective team draws energy and ideas from all members, producing a synergistic result. In total quality, the organization is led through the management chain, but the organization is fueled from the bottom up.

Traditional management theories hold that management's function is one of establishing priorities. The economics of the business is number one. The efficient utilization of workers is seen only as a positive or negative impactor upon that economic priority. This view of management's function has been expanded in the traditional organization to describe the priorities of day-to-day responsibilities. These fall into the order of production, costs, and then quality, very high to high to low, respectively. In a traditional organization, rarely is management held responsible and accountable for worker safety and health. It just isn't an established priority nor is it seen as a significant impactor on the number-one function of management: concentrating on the economics of the business.

Total quality defines management functions and responsibilities much differently. Economics still has to be number one. There is no change there, but how that primary function of management is achieved is what is different in total quality. Accepting the leadership role of management, the function of the business is not *purely* economic. It is more holistic and far-reaching in function and responsibility. Businesses exist not solely to make profit. They must also provide for the economic good of the community, society, and the overall economic structure. In other words, what is economically good in the short-term for an individual business is not necessarily good for the short-term or long-term economics of the community, the society, and the economic structure.

This broadened picture also extends to management responsibility. There must be a balance between holistic responsibilities, including production, quality, costs, morale, safety, and environment. Balance is the key. In total quality, when priorities in one area of responsibility are out of balance, even for just a moment, the total organization loses balance and problems occur. Problems in one area of responsibility always mean problems in the others due to this imbalance.

Total quality management is also more efficient. We use the word "proactive" to describe the way the management function is accomplished. It means to work ahead of need, to plan and execute according to plan rather than by knee-jerk reaction. The opposite, of course, is "reactive." Proactive management is significantly more efficient than reactive management because it avoids problems and streamlines processes.

Participation in Total Quality. Traditionally, there is a wall between management and labor. This wall is defined by differing roles within the organization: those who manage the business and do all the thinking, and those who bring their bodies to work and do the labor. Total quality recognizes that this separation and limitation of roles cannot be successful in today's dynamic and competitive market. Extensive participation, therefore, is a key total quality concept. Ideally, this participation is without limit or division. In a total quality organization, leadership comes from the top but energy and ideas percolate up from the bottom. It is truly a top-down and bottom-up organizational concept.

Two key terms can be applied to this concept: empowerment and ownership. But as in TQM, these words have become more buzz words in American business rather than living concepts. Look at empowerment as a motor standing next to a newly charged battery. Disconnected, as in traditional American workplaces, it sits idle; perhaps it is manually turned over every now and then by management. But if it is connected to the power source, it is energized. In total quality, management begins providing this power by creating a highly participative environment. From that atmosphere of openness, the energy is generated and becomes self-sustaining and ever increasing.

Figure 1.1. Participation in a Total Quality Environment.

Ownership is not financial. It is a deeply personal connection and the source of pride and accomplishment. It is motivating to see that one's labors or thoughts contribute to the success of the organization. Ownership is without jealousy or resentment. Individual and team efforts contribute to the success of the whole organization or business.

In the ideal total quality environment, participation is limitless. We do not, however, live in an ideal world. For this reason, the roles that management and labor play are critical. The leadership role in creating this open, highly participative environment must fall to management. Labor also has a critical role, which, to a great extent, has been defined and patterned by organized labor. Traditionally, labor's role has been adversarial, protective, and one of limiting participation except as narrowly defined by contract. These are challenges to our future that exist due to our history. For example, this limiting of participation is not prevalent in Japan, which has a long history in total quality and organized labor.

Vision, Mission, and Planning. Total quality recognizes that if you don't know why you exist and where you want to go, that any road will get you there. (You also don't know *when* you get there.) Thus, direction and clear purpose are critical elements in a total quality program. Usually, the purpose of an organization or program is defined within a mission statement that merely documents why an element or program exists by defining what and how something is to be accomplished. The statement is fundamental to any element or organization that provides a valued function to the larger organization, business, or society. If there is no value in an organization's existence, it is a waste and should be removed.

A vision is no more than a direction in which an organization or program wants to moves. It is future-oriented, success-oriented, and adds benefit to the program or organization. It identifies a final destination on the road to excellence, toward which a step-by-step sequence is derived to achieve the vision. The plan is this sequence of steps.

Planning can be broken down into three distinctive levels. The strategic level of planning is the highest. Its name derived from military roots, the strategic plan is the master plan, the plan for the entire effort. Strategic plans tend to be more directional than specific. They identify the key targets that must be reached if a vision is to be accomplished. They examine all aspects of the organization or program. At the next level is the medium-range plan, which usually covers three to five years and itemizes specific accomplishments. It breaks each segment of the master plan down into more specific accomplishments. The next level is the annual or short-range plan. At this level, great specificity, such as delivery or completion dates and named accountability, is added to the plan. This is the level at which progress can be affirmed and accomplishment recognized.

Goals, Objectives, and Measurement in Total Quality. The integral parts of the planning process are goals, objectives, and measurement. These obviously are also critical aspects of total quality. Establishing goals and objectives is something we don't do well in America. In safety and health programs, measurement of performance is practically unheard of. We may measure some semiconnected results, such as injury rates, but we seldom measure actual performance.

Traditionally, we set subjective and semiquantitative goals; for example, to "attain or maintain a safe working environment" or "attain or maintain compliance." These goals cannot be measured and attainment is totally subjective; that is, it is up to the perception of others, such as upper management and workers. Too often perceptions of "safe" and "compliant" differ widely. We traditionally set goals for a reduction in injuries and illnesses, calculated by lost time injury rate, recordable injury rate, total injury rate, and lost workdays. By doing so, we assume we are practicing good management. The problem is that we have only marginal (at best) control over achieving these types of goals. If a new doctor who likes a particular type of surgery or likes to write prescriptions moves into town, a department manager lets safety go to pot, or production gets behind, then the safety goal often is forgotten.

In total quality, goal and objective setting and measurement are critically important management tools. But, they must be used correctly to be effective. Actual desired results are the best goals and objectives. Ideally they should be tied to performance, not perception. In total quality, goals, objectives, and measurement form a critical web for achieving success. Too often, what we traditionally use only marks frustration and failure.

Training, Information, and Communication in Total Quality. These three rules of total quality are essential if participation is to occur: 1) Training must precede expectation. 2) Shared information is more powerful than unshared information. 3) Communicate, communicate, communicate.

Training must precede expectation. Total quality organizations are heavily invested in training. Recent surveys of total quality organizations versus traditional American workplaces reveal an 8:1 training ratio. Total quality organizations recognize the critical importance of participation at all levels and that training is the only way to ensure the knowledge and skills necessary for participation. Training is also much more subject-diverse, level-diverse, and individually oriented in total quality organizations. Training is seen as a critical tie to competencies that must be expanded and increased for competing in the ever-expanding global marketplace.

Shared information is more powerful than unshared information. This is the all-too-familiar traditional belief that "information is power" which has historically reinforced the barrier between management and labor in American business. From shared information comes different perceptions, conclusions, and potential input into decision-making processes. We often become myopic in our vision because we want to control information and, thereby, limit the amount of input we receive. We limit our options by controlling information. The decision-making process in turn becomes more subject to error and misdirection. In total quality organizations, information is shared at all levels, input is received and shared, then decisions are made.

Communicate, communicate, communicate. Communication is a two-way concept: I talk, you listen—you talk, I listen. True communication is not natural for us. From our first breath, we talk to others. Depending on our level of need or frustration, we sometimes yell at others. This is not communication. Our ability to communicate is further degraded by our upbringing. "Don't talk to strangers," or "Speak only when spoken too," or "If we want your input, we will ask." Problems with communication are compounded in the workplace by our traditional paradigms concerning two-way communication between management and labor. In short, as a result, in our traditional environment, communication is controlled, devalued, slow, and ineffective. The total quality concept emphasizes an opposite view. Communication is a necessary vehicle for transfer of information. Because communication has many forms, however, effective communication in a diverse, large, or physically separated organization can be challenging. Multiple forms of communication must be regularly used, however, the speed and completeness of what is communicated is of supreme importance in total quality.

Benchmarking and Program Evaluations. Benchmarking has become another buzz word in American business, but it is a very useful and efficient learning tool. Benchmarking is the importation of known successful ideas and practices into a program or organization seeking improvement. Once we learn what actually works, we borrow it rather than using the traditional trial-and-error approach.

Program evaluations are internal tools that accomplish three things. First, they provide a tool for regular and comprehensive program assessment. Second, they document improvement in specific program areas. Third, they identify program areas where improvement can be targeted. Regular assessment is valuable in a total quality program because it assumes nothing and prohibits program myopia.

The documentation of improvement is important because it ensures added value to the organization and gives upper management a comfort zone about a particular program that is unachievable when traditional observational or judgmental management methods are used. Identification of growth areas is critical if a program is to improve and become more effective. It is also an important preactivity to any benchmarking effort.

Continuous Improvement. It is difficult for Americans to understand or accept the concept of continuous improvement because of our lack of patience and our traditional expectation of rapid change. "Innovation" is used to describe the American way and is often equated with other terms such as automation. Continuous improvement (or Kaizen, as it is known in Japan) is a slow, small-step-by-small-step, daily, low-cost or no-cost, everyone- participates concept. This is obviously very different from our usual expectations and processes in America. Kaizen is not a blinding-light concept. It improves a little each day.

The power of continuous improvement is fourfold. First, in American innovation, our planned rapid improvements rarely meet expectations. Kaizen has a 100% success rate. Second, American innovation usually costs a lot of money to implement. Automation, new technology and new facilities are costly. Continuous improvement relies on low- or no-cost improvements. What is spent is spent more wisely, and it has an infinitely greater chance of netting real results. Third, innovation is usually the product of upper management or specialized support functions such as engineering. Significant negative impacts occur when innovations are implemented because they lack the real-world, hands-on evaluation prior to start-up. Once innovations are implemented, they require constant refinement to correct deficiencies, which costs more money. Kaizen relies on ground-floor participation. Because of that, the day-to-day, hands-on elements are accounted for prior to implementation. Because the process is slower, learning along the path is accommodated. Fourth, American innovation is based on immediate results. Such rapid changes lead to poor long-term results, and the inability to respond to performance fluctuations. Kaizen focuses on the *process* from which the results emerge, making it much more consistent, and eliminating large performance fluctuations. Because it is continual, Kaizen has excellent staying power.

ADVANTAGES OF TOTAL QUALITY

The advantages of implementing total quality into a safety and health program are multiple and diverse. From a perspective outside the safety and health program, it simply works better. From an inside perspective, it yields documentable results, is more efficient, and minimizes the traditional problems and frustrations. From either side of the fence, total quality has many advantages. We'll examine the major ones.

More Effective. Total quality simply allows you to get more done. It does so by spreading out the responsibilities and skills so that more people resources can be directed toward safety and health. It also places direct responsibilities, with accountability, to a direct line authority. Correlation of responsibility with day-to-day authority gets the important safety and health job done. This process runs counter to the traditional correlation of more safety and health work equals more safety and health staff. When the responsibilities are spread across the organization, more work can be done with fewer specialized safety and health staff. Total quality is therefore more cost effective for an organization. The resulting greater standardization of the safety and health effort makes across-the-board improvements to effectiveness. The planning aspects of total quality also provide a mechanism for changing to proactive program management which is always more effective than reactive management.

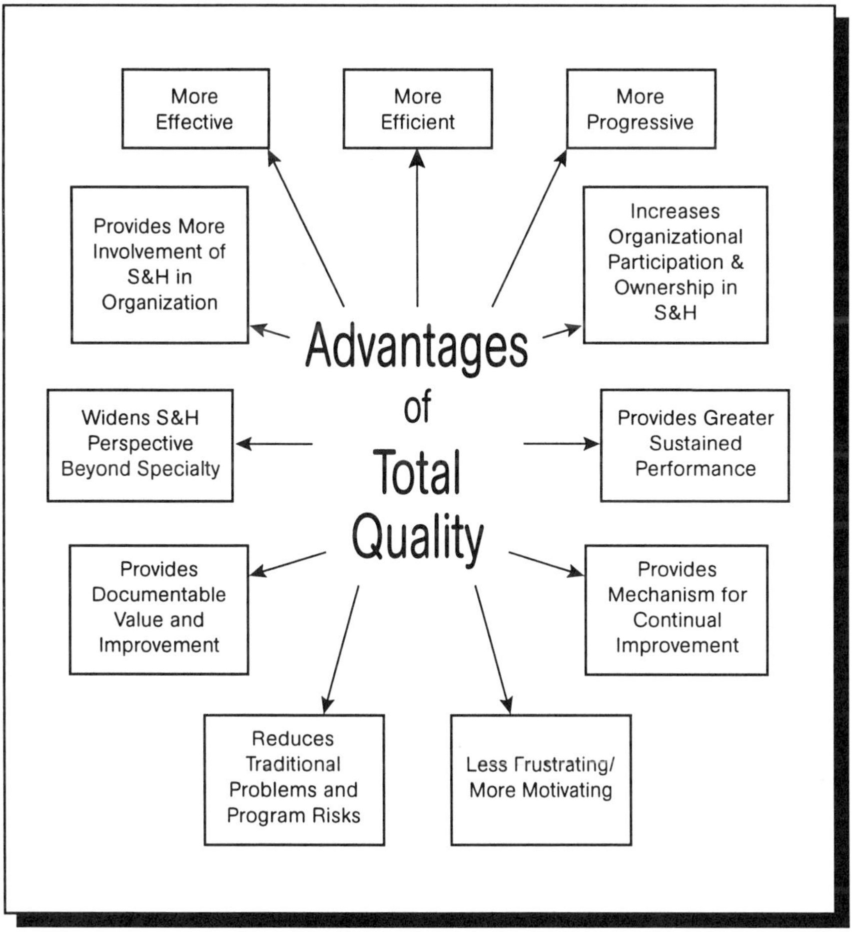

Figure 1.2. Advantages of Total Quality Management.

Is More Efficient. There will never be more than 24 hours in a day. Our traditional role of trying to do it all ourselves cannot work efficiently (1 x 24 = 24). Spreading out the knowledge and skills with designated responsibility and accountability allows more hours to be immediately available (n>1, n x 24 = 24n). Not only is more time available, with more people working toward safety and health day-to-day, but progress is made faster.

Is More Progressive. Total quality is anchored in training and multifunctionality, which increase the knowledge and skill base of all workers. This is contrary to the traditional protection of the status quo by maintaining specialties. Increasing the safety and health knowledge and skills in the line structure also makes the entire effort more progressive from a holistic responsibility approach. Multifunctionality and increased skill level allow quicker changes within the organization to meet specific safety and health challenges.

Increases Organizational Participation and Ownership in Safety and Health. Traditionally, there has been a barrier between participation and ownership in safety and health in American workplaces. Total quality removes these traditional barriers. Everyone participates and the ownership for safety and health is throughout the line structure, from those who make decisions and commit resources to workers who attend to their daily safety needs.

Provides More Involvement of Safety and Health in the Organization. Because total quality is a team concept, with safety and health a part of the total quality team and not a specialized function, the safety and health staff is more involved in all organizational aspects. This improves preventive efforts and allows greater knowledge to be shared directly through these interactions.

Widens Safety and Health Perspective Beyond the Specialty. Too often, safety and health professionals get tunnel vision, seeing only from their specialties and perspectives. This causes isolation and even job loss in downsizing efforts. Safety and health personnel involvement in all aspects of the organization encourages an organization-wide perspective, which is more valuable to the organization. Consequently, decisions cause less problems.

Provides Documentable Value and Improvement. Whose responsibility is it to give us adequate feedback and document our worth to the organization? Traditionally, we would answer, our boss's. The total quality process changes this responsibility. Through measurement of performance and objectives, we can document our own worth and our own improvement. Documentation not only gives us positive feedback and motivates us to continue, it continually tells our boss that we are doing our jobs, providing valuable input to the organization, and don't need to be overly managed.

Provides Mechanism for Continual Improvement. Kaizen is a mechanism for improvement. Using tools like process mapping, program evaluations, benchmarking, time-based management, and partnershipping, we can continually identify where improvements can be made. This mechanism ensures that our programs will be dynamic and not stagnant. Continual improvement ensures us that we will always be valuable to our organization.

Provides Greater Sustained Performance. Total quality is process-oriented, not results-oriented. By working on the process to make it more efficient and effective, to reduce wasted efforts, and to decrease cycle times and waiting time, we can greatly advance results. Moreover, because the process is improved, better results are more constant and less impacted by unanticipated events. The results, therefore, have greater staying power and are more consistent because performance is constantly improving. The presence of a higher level of safety and health knowledge and skills in the line structure also sustains the performance. The peaks and valleys caused by direct or indirect impactors on one or more specialized safety and health staff are removed.

Reduces Traditional Problems and Program Risks. The traditional practice of safety is the role of policing compliance with the safety rules. This has always been perceived negatively throughout an organization. The total quality process places responsibilities where they belong—in the line structure. This removes the need for the safety and health practitioner to be the rules police officer. The safety and health staff is oriented to provide valuable services to the line structure, where and when the services are needed. The change in safety and health role solves many problems that are common in the traditional organization. Because of the change in perception by management, a total quality safety and health program is under less risk in a downsizing effort.

Is Less Frustrating and More Motivating. The traditional way safety and health is practiced is a breeding ground for frustration and low motivation. Safety and health staff have responsibility without authority, constantly have to provide economic justification, must argue cost-benefit relationships for hazard correction, and are at the top of the downsizing list—all of which contribute to frustration and lack of motivation.

Total quality turns this picture upside-down. Responsibility is married to authority in the line structure. Economic justification makes way for documentable value-added services to the line. Cost-benefit of safety and health issues gains considerable momentum when the gains in productivity, quality, and employee participation elements are added to the equation. Since the safety and health staff size meets the service needs of the line structure, the traditional downsizing threat is equalized. Motivation is greatly improved when a program is run by plan and successful performance is measured. Improvements are documentable and add value to the organization.

A professional colleague decided a few years back to change from the traditional approach to managing safety and health, to total quality. He refers to his decision as a "step in faith," much like "stepping out of a fiftieth story building window," as he illustrates it. As he recalls, "All I could see was risk. Risk of changing, but more importantly, risk of not. In our rapidly changing company, we had to change and change quickly if our safety and health program was to survive."

He knew of our program, and we began talking about total quality. The concept merged nicely into the changing organization and culture toward which his company was moving. He took that "leap." I was honored to be a mentor to that change process. I asked him if there was one thing he could share with others, what would it be? He mentioned two things. First, "Change of this magnitude from traditional thought is not for the weak in spirit or those not prepared for the long haul." There was pain, joy, frustration, dismay, exuberance, emotion, tough talk, salesmanship, communication, learning, teamwork, and success. Second, "Believe the press clippings; it's as good as advertised."

CHANGING TO TOTAL QUALITY CONCEPTS

The advantages of total quality over our traditional way of doing things are great. But, this is real life. You may like everything that is in the candy store, but chances are you won't get everything you want. So, if you choose to change to total quality, what can you realistically expect? That depends. It depends on how much you choose to import from total quality and what organizational environment you are bringing it into. For example, you might choose to only bring the planning and program evaluation aspects of total quality into your program. You may feel that these can be easily imported and will improve what you've got. However, if what you're doing isn't working, is increasingly frustrating and demotivating, or is at risk from downsizing, you may choose to totally change how you do your job.

The second impactor to what you can expect from changing to total quality is the way the organization operates. Ideally, total quality can bring maximum advantage to a safety and health program if it is in lock-step with an organization's change effort. Changing to total quality in your program is easier in that the major philosophical and organizational changes brought about by total quality are part of the bigger, organizational change. If, however, you want to make the change to total quality and your organization believes and practices traditional American business concepts, you may not ever get to enjoy all the advantages total quality can bring and a change effort will be more of an up-hill climb.

Table 1.1 compares total quality in two types of organizations. Obviously, total quality has more potential impact in a matching organizational environment. But if you have a traditional top-down American organization, total quality can provide significant advantages and improvements to you and your safety and health program. There is one more common factor. Once something successful is transplanted into a nonchanging organization, it takes root. What works is noticed by others. The mere fact that it works causes other programs to begin to change also. In fact, one program can have a significant impact on changing the culture and organization or an entire business.

Table 1.1. Total Quality in Comparison to a Traditional Management System.

Total Quality Component	In TQ Organization	In Traditional Organization
Line Responsibility for S&H	Merges into TQ Culture	Runs Counter to Tradition
Employee Participation	Merges into TQ Culture	Roadblocks to Participation
Proactive S&H Management	Can Be Highly Proactive	Role Perceptions May Limit Success
Control Programs	Can Be Used in Most Areas	Can Be Used Traditionally
Vision, Mission, and Planning	Can Be Highly Successful	Can Be Highly Effective
Multifunctionality	Can Level-Load Demand	Can Level-Load Demand
Dynamic Rules/Enforcement	High Participation/Dynamic	Low Participation/Traditional
Benchmarking	Can Be Used Effectively	Can Be Used Effectively
Program Evaluations	Can Be Used Effectively	Can Be Used Effectively
Continuous Improvement	Excellent With Measurement	Internal to Program Only

As an example, a safety and health program in a small remote facility of a Fortune 100 corporation decided to follow a different drummer. Against corporate history *and* advice, they chose to use the OSHA Voluntary Protection Program (VPP) guidelines as a program evaluation tool and then, boldly, to apply to the VPP as a benchmarking effort. Needless to say this choice caused a lot of concern and anxiety at the corporate home office. The facility's safety and health program had for some time been a leader in new thinking and its geographical separation from the corporate home office allowed it some latitude. Their effort paid off. They were given Star status under the VPP in their first attempt. As a result, a strange thing happened. The corporation was in search of different ways of doing things. One forward thinking person at the home office took the facility's lead and saw gold: he made a proposal to his boss and then to his boss's boss. He worked his idea up the organizational ladder to the Board and got it approved. That corporation decided to be a leader in safety and health. They set a goal to have all of their more than 100,000 employees covered by the OSHA VPP program or a corporate equivalent that they developed, and set an ambitious timetable to accomplish this goal. Do you think that corporation would have ever reversed its traditional ways of thinking and attained such a progressive goal if a small, forward-thinking program thousands of miles away hadn't done it first and showed them the way? I don't think so. Success can be contagious.

CHANGE WITHIN SIMILAR ORGANIZATIONAL EFFORTS

Let's make one differentiation first. For these principles to apply, the organization has to be changing to total quality, not something called total quality that resembles a reincarnation of traditional management practice. The organization's change doesn't even have to be formally called "total quality," but the change must be fundamental and the components and concepts must match. American efforts toward total quality are often in name only, and are halfhearted or resemble downsizing more than total quality. When total quality is implemented into a safety and health program in such a false change environment, it resembles a more traditional assault rather than a true total quality effort.

It's exciting and exhilarating to see an organization making a positive change with a well-organized effort. Resistance to change is natural and seek a safe harbor (where change is not happening) is natural in a changing environment. When everything is changing in the same direction at the same rate (more or less), there is no hiding place. Resistance is more directly confronted and dealt with at one time rather than via the normal practice of guerrilla warfare throughout the ranks. Total quality changes in a safety and health program have the highest rewards and the easiest pathway to them is this lock-step changing environment. Many of the concepts of total quality, such as participation and line responsibility, are negotiated and resolved at appropriate levels in the organization. Normally, these levels are well outside and above safety and health programs. Having the large battles fought and won by upper management makes these aspects of total quality attainable.

It is also more difficult in this lock-step environment for any one area or program to eat half the apple. Generally, when the vision is for total change, all areas are swept into the change stream so all aspects of total quality are adopted and become part of the organizational fabric. All threads become part of the total fabric. All threads keep the organization tight and give it the intended pattern, depth, and texture. The act of pulling one thread from the fabric loosens the weave, makes the fabric weaker, destroys the intended pattern and can notably change the texture. Each component is dependent on other components and everything becomes an integral part of the whole. This is an organizational total quality effort.

CHANGE IN A TRADITIONAL AMERICAN ORGANIZATION

Let's be realistic. Making a complete and well-organized change is rare in American business. The vast majority of the time, a unified change is out of reach. More often than not, you are out there on your own. This is why this reality check is important for us to talk about. What can you realistically expect if you choose to change to total quality in a traditional American organization? In this environment, you can choose to take certain bites of the apple. You cannot eat it all. Knowing this, the resulting benefits are greatly dependent upon how many bites you can take, and in what order and how quickly you consume the portions you choose. There is also much you can learn from other such efforts to help improve your successes and decrease your level of effort.

When implementing total quality into a part of a traditional organization, a simple rule should be followed: Begin inside before you move out. The following table itemizes the opportunities for improvement in a traditional organization.

Your opportunities exist in all but two areas: line management responsibility for safety and health, and dynamic rules and enforcement. The remaining areas have opportunities for improvement with total quality. Arguably, some components are significantly hampered without the benefit of an organizational total quality effort. But these reductions in impact, to a large extent, can be minimized with good program leadership and building of partnerships outside your program.

In what order should the components be implemented? Experience shows us that the following order is the most successful.

Table 1.2. Opportunities for Total Quality Improvement in a Traditional Organization.

Total Quality Component	Inside Opportunity
Line Responsibility for S&H	Won't happen!
Employee Participation	Your Staff Only
Proactive S&H Management	Too Dependent on Outside
Control Programs	As Traditionally Defined
Vision, Mission, and Planning	Great Opportunity!
Multifunctionality	Great Opportunity!
Dynamic Rules/Enforcement	Won't happen!
Benchmarking	With Management's Blessing
Program Evaluations	Great Opportunity!
Continuous Improvement	Great Inside Your Program

Staff participation is an obvious first step. It is usually under a safety and health program manager's complete control and participation is the key to total quality. Providing a destination (change vision) and determining why your program and functions exist, and planning to succeed comes next. This is an opportunity to begin building multifunctionality in your staff, but it is a real challenge to paradigms that can be impacted through training. Program evaluations give you a snapshot of what your program looks like, where it's strengths lie, and where improvement is available. It is a tool for building a stronger safety and health program. Control programs have been available to safety and health programs for a long time. But, when a program and staff begin to think differently, it opens up opportunities for control programs that were previously unseen. In turn, all the previous components of total quality can't help but move the program toward proactive management and away from the traditional reactive style. The concept of continuous improvement then benefits directly from participation and program evaluations. Improvements are incorporated into the planning process and fuel more participation and improved or increased control program opportunities. Finally, benchmarking becomes available as a tool from which to learn success. This, in turn fuels the continuous improvement engine of total quality.

Most Successful Implementation Order

1. Staff Participation
2. Vision, Mission, and Planning
3. Multifunctionality
4. Program Evaluations
5. Control Programs
6. Proactive Safety and Health Management
7. Continuous Improvement
8. Benchmarking

Total quality can be an effective concept for improving any safety and health program. It can be applied internally into a program seated in a traditional American organization or can be part of an organization-wide total quality effort. The power and advantages it brings to a safety and health program are great and diverse. Although it is best if part of an organization's total quality effort, it can turn a traditional safety and health program around and have a tremendous impact on worker safety and health. The critical steps are making the choice to change, deciding how much change you want to ultimately make, and working through the change process for successful implementation.

2

The Dynamic Process of Change

Early in my career, I was actively courted by another company. The new job would have required that my family move across the country. One night after dinner I opened the question for discussion, "Anyone want to move East?" A wide and interesting range responses immediately came back. My youngest looked worried and just looked at his Mom for her response. My next-oldest son was obviously pitching the idea around in his head and responded, "Yeah, there's a lot of history and things to see there." My oldest, without looking up from his homework just responded, "Cool!" My wife was exactly that—cool, very cool. "You're not serious are you?" Getting up from the table, she continued, "Well, this dinner is going to have a hard time digesting now."

These reactions to a potential move are probably typical, but why were the responses so different? The responses are fairly predictable and are correlated to the age of the respondent. We all deal differently with change, especially unanticipated change. The very young mirror the reactions of their parents. They learn by observations: "If you're worried about it, perhaps I should be also." As kids become more independent, they look at change as merely another horizon to look beyond. Adventure is never unwelcome. As we mature, most of us lose our desire to explore. We tend to cling to what we know and can depend on. Change becomes a foe rather than an exciting opportunity.

Change is a frightening thing to adults. We like predictability and change is anything but predictable. Imagine coming home after a long day at work, opening the front door, and finding different furniture in your living room, a strange dog greeting you at the door, and an unfamiliar child asking, "Who are you?" Jimmy Stewart, in the classic motion picture *It's a Wonderful Life*, experienced such a loss of predictability. At a very low point in the character's life, he wishes he had never been born. Magically, it becomes so, yet he is still there as an outsider witnessing the changes brought on by his wish. His reaction to the loss of predictability is fear. To compensate for his fear, he begins a desperate and frustrating effort to regain predictability.

Change is frightening because it disrupts the predictability that we depend on. The way change is implemented can also have a huge positive or negative impact on our reaction to change. Sudden change and change that occurs without our control has a magnified, disruptive impact and can be frustrating and demoralizing. Change that is carefully planned, prepared, and implemented, however, has a greatly reduced negative effect on those impacted by the change. It can also significantly increase the chances for, and reduce the time required for, implementing a successful change. These are important concepts for us to remember as we explore the process of change.

COUNTERMEASURES TO REACTIONS TO CHANGE

Change is inevitable and unavoidable. Are we sentenced to a life of fear and frustration because of this? No, because how change is accomplished is critical to our reactions. Some important aspects of accomplishing change can notably decrease our fears and frustrations. They are vision, benefit, planning, communication, and participation.

Vision. In a change process, we need to know where we are going. It is very unsettling to be embroiled in change for the sake of change only. Imagine writing a proposal for your boss. He reads the proposal, and says, "It's not quite there yet. You need to make more changes." "What changes do you suggest," you ask. "I don't know, but I'll know when it's right." Can you imagine the frustration, confusion, and inner stress that you would feel? Without knowing where you are going, you haven't a road map or a destination.

Lou Gerstner, the CEO of IBM, stated that, in his opinion, vision was overrated. Therefore, he set out to change Big Blue without a vision (or without a clue, as most economic analysts have said). In what area did IBM gain it's tremendous market advantage? It did so in the mainframe computer lines. Where is the market expanding and growing the greatest, mainframes or PCs? Along with some deep reductions in force to solve financial problems, IBM began expanding their PC offerings, quality, and price. Soon though, the natural pecking order took over. The PC areas were devalued and the mainframe direction gained strength. In short, today IBM looks almost like it did 10 years ago—a little skinnier, but the same. It certainly is not poised to recapture the world computer market. Solely because it is in the computer market, it will have years of high profits and others of large losses. What does this example tell you about the importance of having a vision when implementing change?

Benefit. There must be a reason for change. Normally the reason centers around some benefit to the organization or program. Perhaps if the business doesn't change, it won't succeed. Maybe the present organization is not working or the current results aren't what you or your leadership wants. Ideally, the benefit should be mutual to those involved in the change. For example, if we are successful in this change, you can keep your job or get a new, more challenging assignment, or the business continues to operate. Imagine the results of a massive change because the head of an organization merely feels like revitalization is needed. Ninety-nine percent of the time, changes powered like that will fail during the initial stages.

Have you ever tried to get your spouse (or significant other) to lose weight or begin an exercise program? How did it work out? Chances are that it didn't because there was a lack of mutual benefit.

Planning. If you know where you are going (vision) and why it's important to get there (benefit), you need to plot an effective strategy to get you there. That's planning. You need some sort of point-by-point plan, including a detailed road map of the journey, before you can start. Without one, it would be like making a commitment to drive from Miami to Seattle without an assigned route and without knowing when each milestone should be accomplished. Mathematically, there are an infinite number of routes including some that run parallel to or away from the intended target. If you took a meandering journey in such a manner, it is highly doubtful that you would ever get to Seattle, and assuredly not on time.

Think about a plan in this way. Let's say that you want to shop at a new specialty store across town. You get the address from the advertisement. As you get into your car, you draw a mental map of how to get there: what roads to use and what turns to make. When you get close, you slow down so that you won't miss the turn into the store. The store (with address) is your vision. What you will shop for and perhaps buy is your benefit. Your mental route map to the store is your plan. You would not arrive if you didn't think anything about the route you would take.

Communication. Successful change is all about communication. It must be constant, consistent, informative, up-beat (as possible), honest (above all), and repetitious. Communication begins long before the actual change effort and continues long after the change is successfully implemented.

Can you imagine a trip in which you don't speak to your traveling partner before you leave, during the trip, or after you get there? You might as well be alone. Have you ever heard the phrase, "The silence was deafening?" If you stop communicating, those involved in the change feel abandoned. Those impacted by the change will naturally assume that "No news is bad news." Lack of communication causes a lot of unnecessary anxiety and can derail any change.

Participation. If you want the best chance for successful change implementation, participation is a critical element. It's a simple mathematical correlation. The more people you have involved in a change, the better the chance of success. In addition, the more input you have before and throughout the change, the fewer setbacks you will have along the way. How do you get participation? It is often as easy as asking for it. You won't always have positive participation. However, even nonpositive participation can be good, because it helps prevent covert actions and negative communication, which are disruptive to change.

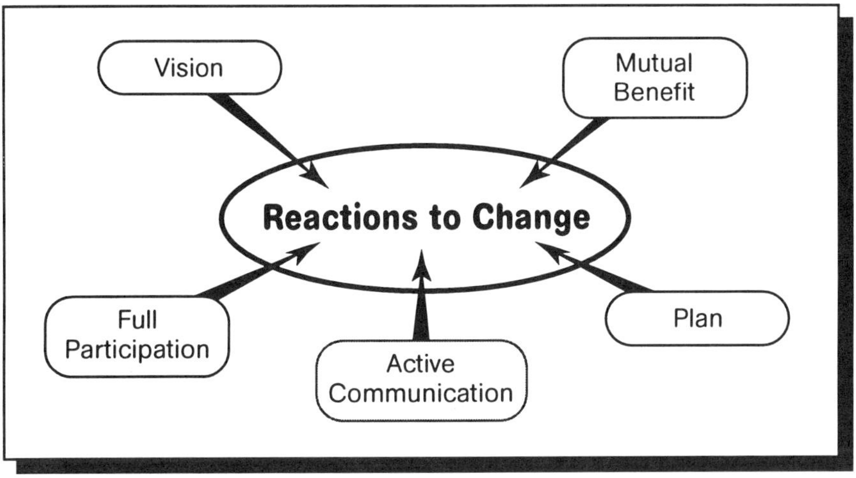

Figure 2.1. Participation Is Necessary for Change.

WARNINGS

Change and disruption are nearly synonymous. Change never happens in the blink of an eye. Time and patience are required. This is an important consideration because we Americans aren't known for our patience. Change requires that you stay the course and maintain a positive outlook. Don't start the process if you don't have the patience and the resilience to see it through to the end. Short term changes are commonly known as the "plan of the week." Those impacted by a short-term change effort will naturally have a "this too will pass" commitment and low participation in that effort.

A number of warnings are appropriate and critical for us to heed before we begin.

Stay the Course. Change is like a long ocean passage. You must be prepared to steer in one direction for a long time before you can expect to see your destination on the horizon. Changes in course only disrupt the voyage and delay the outcome. If you aren't willing to invest time and stay at the helm, don't start in the first place. This challenge is confounded by the magnitude of the change and the number of people affected. The larger the change from the traditional and the more people impacted by it, the greater the amount of time required to implement that change. In extreme and rare cases, huge changes can be accomplished overnight, but these usually involve change-or-die scenarios.

Patience Is Critical. I know of no other process, except maybe marriage and raising kids, where patience is more critically needed. Webster defines patience as the will or ability to wait or endure without complaint. We must expect the time element associated with change, expect problems along the way, overcome resistance, continue the effort until it is complete, and not whine about all that is required of us. In particular, those leading a change should refrain from complaining. After all, leaders become role models for those impacted by the change. If a leader buckles under the pressure, the followers will, too.

Patience is a difficult attribute to acquire. If a change is needed, our response is, "Let's get on with it, now." This is the way we deal with our kids, car repairs, weight loss, and a multitude of other things in our lives. Lack of patience is one of the major reasons why Americans compile an impressive amount of debt on credit cards. We simply do not have the patience to wait for the cash. My son came home the other day with a new $500 CD player. Did he need it? No—he has a less expensive unit in his room. Could he afford it? Of course—he had his credit card. He is typical of many of us.

Resistance Is Normal. It makes little difference if the change is positive or negative. Change is disruptive because it affects predictability. Disruption is always met with resistance. Resistance, therefore, is normal, and can be dealt with positively or negatively. Dealt with negatively, resistance temporarily goes underground. It's like a pressurized container that leaks to the surface gradually and at several points. You spend all your time sticking your fingers in the holes. If, however, it leaks one place all at once, you have an unmanageable gusher. Resistance should be dealt with positively and kept out in the open.

One of my sons is "neatness challenged," to use today's jargon. His part-time job requires that he wear a white shirt and tie which end up on his bedroom floor. His sloppiness adds to my wife's washing and ironing chores. There seems to be a lack of commitment on my son's part to hang up his work clothes. My wife had hooks put on the side of his bookcase upon which he could hang his clothes. He then hung his jacket, a belt, and his hats on these hooks. His work clothes remained on the floor. Why did he resist this simple and convenient change? Because the hooks provided a new place to hang things for which he had no place before. He put the work clothes on his floor and that was predictable. He was patterned and resisted the change. What happened? His Mom was able to convince him in a rather heated fashion to make the change. He had to choose between the predictability of his clothes being on the floor or a peaceful relationship with his Mom.

Perspectives Differ. It's a simple fact that everyone doesn't see things the same way you do. Perspective comes from our frame of reference (FOR). Each of us looks at the world through our own eyes, with our own emotions, needs, and fears. This is our frame of reference. Our FOR in a change effort would be, "What's in it for me?" If no value is seen, we are not interested. If risk is seen, resistance is natural. If no benefit is seen, roadblocks or lackadaisical efforts will be the result. If threat is perceived, anger can be expected. Perspectives are not good or bad, or right or wrong. They are realities framed by each of our FORs. Herein lies an opportunity for communication and participation in the process.

A manager wanted an operator of a portable grinder to change tools. The new tool had a better guard and less vibration. It weighed more, which would mean that the operator would have to apply less pressure on the grinding surfaces. The manager's FOR was safety, prevention of cumulative trauma disorders, and work effort. The operator's FOR was different. He saw extra weight that he would have to lift, more bulk that required more effort for him to use, and he had never been bothered by vibration like some of the other operators had. Because of the different FORs, the new grinder was purchased but not used. Too often we ignore or devalue the FOR of other's, thinking or hoping that they will eventually see the wisdom of our decision.

Know Thyself. Before you start change, you had better know yourself well: where you are vulnerable, what you fear, where your strengths and weaknesses lie, what skills and knowledge you need to develop, etc. The middle of change effort is a poor time to find out something about yourself, and an experience that can derail the entire change process.

Know Thy Process. Too often we set a direction for change and really don't know if it will make things better. Most of the time it's because we don't take the time to know the process that we are changing. Consequently, the change has difficulty gaining support from those who know our process better than we do. If the change is implemented, they know it won't help.

For example, a furnace was going to be upgraded and $250,000 was set aside. This furnace oxidizes materials on roll-out and roll-in, rail-mounted carts. Part of this change effort involved spending considerable money to expand the number of carts so that more material could be oxidized by the furnace. It was thought that this approach would notably increase the effective utilization of the furnace. In the first meeting to discuss the changes, it was obvious that there were differing opinions on how to increase the use of the furnace and about cart design. None of the options could be discounted. A team was appointed to detail the process using a technique known as process mapping. In a little over an hour, the team got to know the process from input feed streams to product completion. From their new-found knowledge and the level of increased production needed, the team discovered that the existing cart number and design would accomplish the new production level with minor process modifications. The upgrade was accomplished for under $50,000.

Making a decision to change is easy. Throughout our lives we have done it many times. Traditionally, we call them New Year's Resolutions. We resolve to quit smoking, exercise more, be less critical, lose weight, etc. What's our success rate? Not good, because implementing change is anything but easy. For this reason, navigating change is not for the weak of heart. It takes a reason with a lot of gravity. Seldom is just having the opportunity to change enough to get us truly committed. Don't go into change without recognizing the realities of the commitment needed and the stresses that will be created. The degree of each is directly proportionate to the amount of disruption the change brings. Success, to a great extent, is dependent on how important the change is.

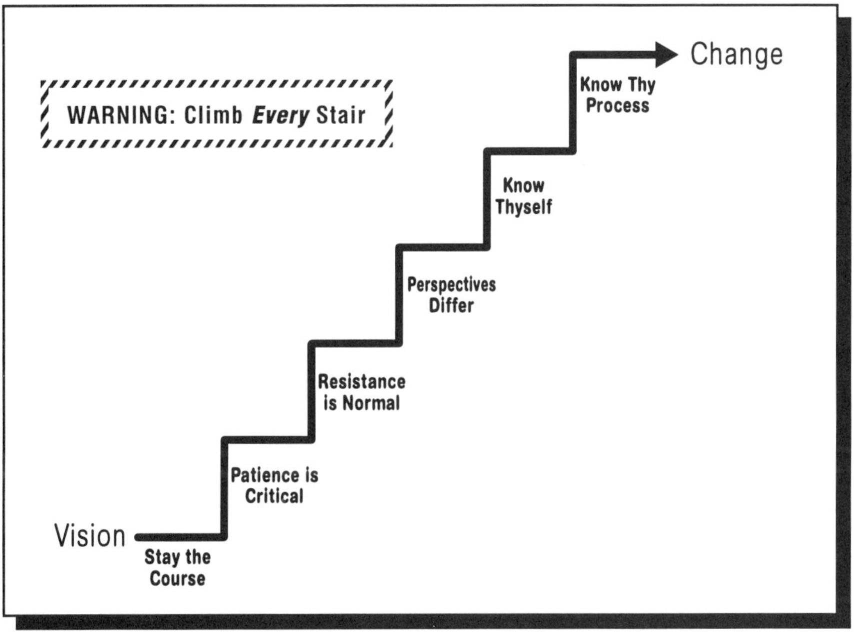

Figure 2.2. Steps Toward Implementing Change.

PITFALLS

Important pitfalls are those aspects which, if not avoided, can significantly disturb your change process, add a lot of frustration to the effort, and significantly impact the chances for successful implementation. In our discussion of pitfalls in the change process, it is appropriate that we introduce a flow diagram illustrating the questioning process that should precede any change. This questioning process is predictable. It is the process we choose that is uncertain and the end result of change that evades our vision.

The purpose of this pre-change questioning process is to identify pitfalls that exist and make them visible. When they are visible, we can rethink or modify the change process to avoid these pitfalls. Let's briefly discuss these pitfalls and how they can derail change.

Pitfall #1: Change in a Nonchanging Environment. When a friend of mine and his wife started a family, they decided that one parent needed to stay at home. Since he had established a career path and her less career oriented, they decided that she should become a homemaker. For many years he continued his job supporting the family, and she stayed at home and cared for three children. One day a few years ago, she announced that she had secured a job and would be returning to work. A change of that magnitude sent some significant shockwaves through the family unit. They hadn't taken the time to talk about the change before it was announced. This oversight was magnified by the fact that change was occurring in an environment that hadn't changed for quite some time. The home had become highly predictable. When significant changes in a program are to be implemented within a nonchanging organizational environment, a similar situation can exist. Generally, communication in a nonchanging organization is poor if it exists at all. Any change is seen as disruption to a long-standing state of predictability.

This certainly does not mean that if your program exists within a slumbering dinosaur, change is not practical. Quite the opposite. Often times the beast can be awakened by the benefits of change. But the initial change, the assault to the status quo, will be met with significant challenges. If you don't anticipate the challenges and prepare for them, they can prove formidable and drive the change into the ditch.

Pitfall #2: Change for the Sake of Change. There must be a benefit derived from the change process. Changing the way you manage, the way a program functions, or how a group works is not like changing hair color. If you don't like your hair color, you can change it quickly, and just as easily change back. Change to programs or organizations is not so simple. The idea is to stay the long course for implementing the change and to see an improvement once you achieve the metamorphosis. It is long-range and there must be benefit once achieved. Change, therefore, without clear and reasonable chance of benefit, will run onto the rocks quickly after the journey progresses.

Pitfall #3: Changing a Lot When a Little Will Do. Change can be like cleaning out your closet. If you are like most, you tend to accumulate clutter and old clothes until there is no room for anything else. So, you begin by taking everything out. You vacuum the carpet and dust the shelves. Then you discard the "junk" as you rearrange the remainder. Voila! Change! You then begin the cycle of reaccumulation.

This example can be related to a common problem in traditional American business. When change is necessary (usually downsizing), management hacks away until "corporate anorexia" is reached. Then, being too lean, the organization must expand to accomplish the necessary functions or to keep from losing customers. This example, like your closet, is cyclical. Both are a function of this third pitfall—too much sudden change when a little or more gradual change would have been much better. The end result? Sudden and drastic change tends to be cyclical—if the anorexic state doesn't go too far—and short lived.

Pitfall #4: Change that Conflicts with Larger Organizational Changes. Larger organizational change can be a springboard or an anchor to change within a program. If the change methods or direction agree strongly (or appear to agree), there is a natural greased slide for implementing your change. Many of the organizational barriers will be swept away in the larger organizational change effort. If, however, your method of change or end result is at odds, your road will be steeper and there will be more and higher roadblocks. This does not mean that you should not initiate beneficial program changes when this mismatch occurs. But you should be as prepared as possible for roadblocks. Interorganizational changes that run counter to larger organizational change efforts are more difficult to implement and to get support for; they require much more effort and meet with much more resistance, and have a greater chance of failure. In short, you must be very committed to the change and have superior leadership and communication skills.

Pitfall #5: Change that Conflicts with the Organizational Culture. Culture is the unwritten rules of an organization which lie deeper, are more complex, and are more restrictive than the official rules of the organization. If your desired change result does not complement your organization's culture, you will need strong sponsorship from high in the organization to have a chance at success. Cultures change slowly and require a long, sustained effort from high-level sponsors. Culture conflicts can be one of the largest pitfalls to the change effort. Lou Gerstner at IBM can attest to that.

Pitfall #6: Change that You Don't Really Want to Make. Change requires a high level of commitment on your part. This is one of the reasons why orders to change from high up often just fade away in an organization. The change agents have no "fire" for the change. For example, if your spouse wants you to lose weight, but you might not be committed to the change. However, if you are experiencing medical problems and your doctor tells you to lose weight, you are more likely to become committed to accomplishing that change. Your commitment to change cannot be swept aside as unimportant. If change is needed or desired in your program, you must be committed to that effort.

CRITICAL PHASES OF CHANGE

At one time, I was helping a company struggle with some changes in their organization. The business had been going through significant changes in order to survive in their market. The workers were disgruntled and highly suspicious of management. Line management complied with the changes, but actively grumbled. I was brought in to help smooth things out. Within my first five minutes onsite, I knew what the problem was. I asked, "How was the last change implemented?" My company contact said, "The Chairman said do it so we did." Everyone else in the room seemed to nod on cue. "How was it communicated?" I asked. "Do it now or we'll get someone else who will," was the somber reply. "What do you need my help for?" I asked. "You already know the problem."

Most change is accomplished poorly. It's little wonder that it so often fails or sends cataclysmic shockwaves through the organization. After all, change is full of uncertainties. You might know the direction you want to go (vision) and have already made the decision to proceed, but it is critical that you determine the most efficient path to take. This is most often the faltering point in a change effort.

As with any voyage, uncertainties abound. Most uncertainties can be dealt with by preparing for each eventuality. We can take some comfort because the change process is also fairly well understood. The change implementation process moves through several predictable stages. Each is critical to successful implementation. Each must be planned for and accomplished, in order to progress toward success. There are six stages.

The Stages of Successful Change

1. Evaluation and Preparation
2. Solicitation and Creation of a Change Environment
3. Planning and Implementation
4. Transition
5. Celebration
6. Continuous Improvement

Each stage is important and must not be overlooked or bypassed. Because we are impatient, we tend to skip the first two stages and begin with planning or skip straight to implementation. Perhaps we get together with some key people and agree that the change is a good idea and then just "get on with it." This approach is cluttered with the bones of unsuccessful changes all along its path. It is also cluttered with the would-be careers of people who so unwisely assumed that change was merely getting from Point A to Point B.

It is natural for us to discount the last two stages also. Once you have "arrived," what else is there? These too are important, not as necessary to the implemented change as the first two, but critically important to any future change you might want to make.

We can divide the stages of the change process into three groupings: pre-change stages (the first two stages), change activity stages (the next two stages), and post-change stages (the last two stages).

Each grouping has it's own needs, contributions, and pitfalls in the change process.

PRE-CHANGE STAGES

Evaluation and Preparation. No runner would consider entering a marathon without careful evaluation of and preparation for that event. One cannot just show up at the starting line of a marathon and hope to finish the race without these two preliminary steps. For a runner, evaluation would include both mental and physical assessments. A marathon, like change, requires a high level of mental toughness and conditioning. Mental status and physical readiness must be known quantities. Long races require training that is targeted at meeting specific mental and physical needs. That's where the preparation stage comes in. Training must be specific and gradual for the runner to be ready, both mentally and physically, on race day.

Such is the case for implementing successful change. The first phase of the change effort is very similar to running a marathon. The difference

Evaluation and Preparation

is the number of people involved and the amount of time each takes. Marathons end much quicker. If you have made the decision to make a change, you must begin with an evaluation. There are many areas to evaluate including yourself, your change team, the key supporters needed to implement the change, the organization, the culture, the history of past changes (positive or negative, successful or not), and the change methods you wish to use. Evaluation is a detailed look at the entire environment and players.

Solicitation and Creation of a Change Environment. Change is a team dynamic. It involves those who are actively a part of the change and anyone whose normal responsibilities or processes are impacted by the change. Change must also be authorized by someone. How high up the management chain the authorization comes from is directly proportional to the size and the focus of the change. Authorizations must be solicited and maintained if a change is to be successful, much less get off the ground. Likewise, an environment that can support the change effort must be created. This is similar to stirring the pot, letting the sediments of the organization or process rise to the top and then letting them resettle in a complementary fashion. This stage requires searching

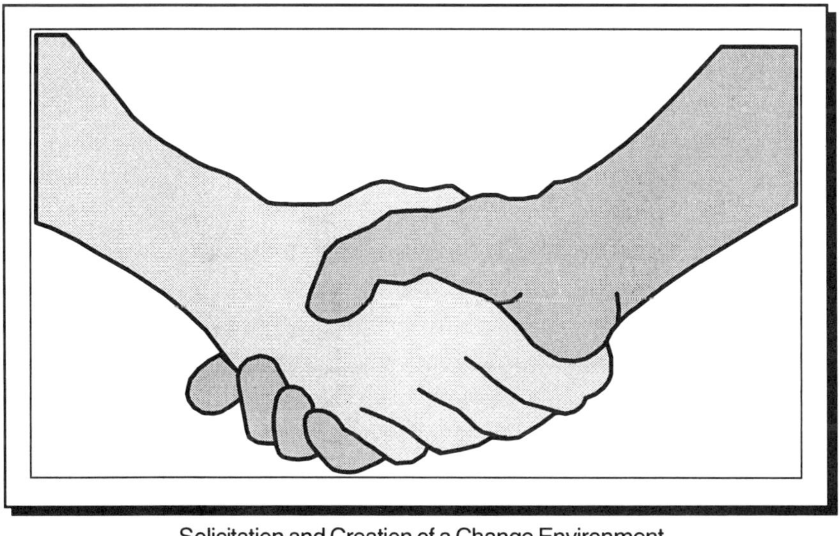

Solicitation and Creation of a Change Environment

for one's purpose, identifying skill or knowledge needs, spreading the change vision, communicating well and thoroughly, and listening actively. It is an important stage, not only for getting your ideas approved, but also for preparing the field upon which your change will "grow."

CHANGE ACTIVITY STAGES

Planning and Implementation. Most of us are poor planners. If we want to accomplish something, we tend to just charge into it and try to get it done. For example, suppose we want to change an accident investigation form or industrial hygiene database format. Too often we go ahead and change it and let everyone catch up with the new way. This approach creates tension, animosity, and frustration. Planning is no more than writing the orchestral arrangement before you play it. In contrast, the end results of changes that are implemented in a free-form approach seldom resemble what was envisioned in the first place. Like a football team that prepares for the big game, to be successful, we must plan what we are going to do and ensure that all players have the skills necessary to execute the game plan. This is a team-based, interactive stage. There is a lot of evaluating, selling, communicating, listening, and rearranging.

Some divide planning and implementation into two separate stages. At first look, the stages seem to be separated by the firing of a starting gun—when the actual change effort begins. However, when one realizes that the change effort actually began when it was obvious that the current way was not working, the starting gun differentiation between planning and implementation no longer applies. In reality, it is nearly impossible to determine when planning ends and implementation begins. Planning is an on-going, dynamic process that is adjusted according to feedback and input, and adjusted accordingly. Planning is not a "set in cement" process at any time in the change process. It cannot be. The implementation plan is constantly changing and this is reflected in the actions carried out in the actual implementation.

Transition. Runners have a saying, "The most difficult part of a race is the middle." Having the start and finish so far away, it is most

Planning and Implementation

Transition

difficult to keep focused on the strategy and to ignore the mental and physical stress-caused complaints in the middle of a race. Similarly, transition is the most difficult and perhaps the most critical stage of change. It is the time when resistance is at a maximum and support is potentially in a lull or in decline. It is the period when communicating and continuing the constant effort are most important but are also difficult. It is also when the vision of the end result is the dimmest and we are the weariest. Because of this, transition must be planned for and active support mechanisms built into the process. Transition can never be avoided. Like an ocean to be sailed across, it can either be successfully traversed or be responsible for sinking the boat.

POST-CHANGE STAGES

Celebration. Celebrating accomplishments is not easy for us. We tend to take things in stride and just move on to the next thing. We need a reason to celebrate. Why else would there be holidays? If we didn't have a Presidents' Day, would we still celebrate our nation's past chief executives? If there wasn't a Martin Luther King Jr. Day, would we still stop and celebrate a great American's impact? Of course not. We need reasons. We need scheduled occurrences for us to take the time to celebrate. For example, everyday is really a birthday but we only celebrate life on the anniversary day of birth each year as society has dictated as the day for celebration.

Celebrating is often overlooked as an indulgent and meaningless excuse to party. Not so! Celebrating is critically important to change because it signifies when the race has been won. It rejuvenates us for the road ahead. It helps us to gain perspective and measure our effort with our past and future. Celebrating is a critical stage in change.

Continuous Improvement. This is one of the most confusing stages to Americans. Continuous improvement was derived in Japan and is a concept by which bit-by-bit and step-by-step, something is improved every day—definitely a long term concept. Americans tend to want to make large-scale, quantum-leap changes. "Why mess around with the little stuff?" we argue. The Japanese, coming back from near-

Celebration

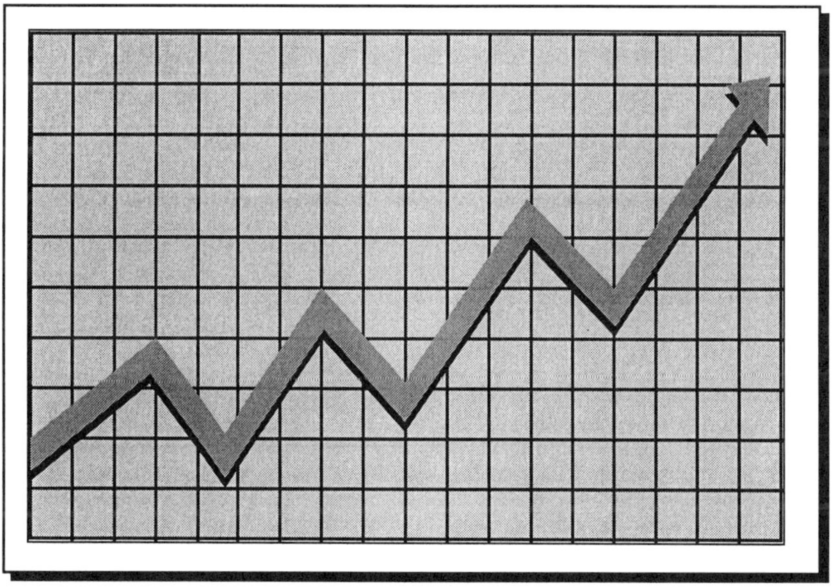

Continuous Improvement

total destruction following World War II, approached improvement from a different direction. They felt that they really didn't have the resources or the knowledge to suddenly get better. Continuous improvement was the only method available. It worked. Continuous improvement places emphasis on a fair understanding of your processes, a high level of participation in the change process, and measurement of progress. Continuous improvement is also an accepted long-term concept that continues after a change is successfully implemented. If it is done well, it ensures that the true sum of all the parts equal more than the actual arithmetic sum. Furthermore, the sum continues to grow, much like a savings plan.

DO YOU REALLY WANT TO DO THIS?

This was the last of the pre-change questions. Our traditional stance to change is what's in question. For example, New Year's is a traditional time for making resolutions: our traditional and accepted way of implementing change. Change, of course, comes from our knowing that a change is needed. What percentage of your New Year's resolutions are successfully implemented? If you're like the vast majority of us, the percentage would be very small. If we have such an unsuccessful track record for resolutions, why do we make them every year? Well, it's traditional! New Year's is that time we have set aside to examine what we have accomplished (or not) and who we are or have become, and decide to better the process or atone for past failures. Does this tradition somehow have any impact on our commitment to the resolutions we make? Quite the contrary. We actually *don't* anticipate accomplishing most of what we resolve. It's kind of a wish list, a formal atonement, a ritual of cleansing.

This empty resolution is so widespread in our personal world, it is natural that we make change resolutions in other parts of our lives, including work, and don't expect to really get them done. That's why this question, "Do you really want to do this," is important for us to formally consider. In reality, the decision to change only counts if we are serious and committed to getting it done.

MESSAGES FROM AMERICAN BUSINESS

Jack Welch, the CEO of General Electric, is one of the most respected people in the business world. Under his leadership, GE has become one of the leading businesses in America. In creating this front-running corporation, Jack Welch based his and his company's actions on certain personal beliefs. Documented in Robert Slater's book, *Get Better or Get Beaten: 31 Leadership Secrets from GE's Jack Welch*, these ideas show us dominant business-guiding principles that are common in the reshaping of American business. What can Jack Welch's

Jack Welsh's Leadership Secrets for Business

- Change, Before It's Too Late
- Look Reality in the Eye and Don't Flinch
- Be Ready and Eager to Rewrite Your Agenda
- Managing Less is Managing Better
- In Deciding How to Change Your Business, Nothing Should be Sacred
- Fix, Close, or Sell
- Don't Get Stuck in the Past—Be Open to Change
- Reexamine Your Agenda Constantly and, if Necessary, Rewrite It
- De-layer: Get Rid of the Fat
- Express a Vision, Then Let Your Employees Implement It on Their Own
- Get Faster
- Remove the Boundaries
- Search for Synergies Between Your Businesses—Strive for Integrated Diversity
- Empower Your Workers
- Eliminate Unnecessary Work
- Aim for Speed, Simplicity, and Self-Confidence

How We Can Be Successful in a Changing Business Environment

- We Had Better Change, Before It's Too Late
- Quit Circling the Wagons of Acquired Turf and Traditional Practice
- Be Ready and Eager to Rewrite Your Agenda or Someone Else Will
- Letting Go of Traditional Practices is Managing Better
- In Deciding How to Change Your Program, Nothing Should be Sacred
- Get Involved or Get Out
- Don't Get Stuck in the Past—Be Open to Change Because It *Can't* Be The Same
- Reexamine Your Agenda Constantly and, if Necessary, Rewrite It, NOW
- De-layer: Get Rid of the Wasted and Redundant Efforts/Approvals/Etc.
- Once a Vision is Expressed, Get Involved Implementing It
- Get Faster at Everything You Do
- Don't Become a Boundary or Limitation
- Search for Synergies in Your Organization — Work Toward Integration
- Empower Workers to Take Ownership in Safety and Health
- Evaluate Your Processes and Eliminate Unnecessary Work
- Aim for Speed, Simplicity, and Self-Confidence in Everyone

business beliefs tell us in our search for successful change? Let's look at 16 of the more prevalent ones and see what wisdom they provide.

Welsh's ideas are from a CEO's orientation, which recognizes the need and causes the change to come about. Our orientation as safety and health practitioners is different. Generally, we are the ones acted on by those changes. For clarity's sake, we paraphrase Jack Welsh's ideas into reflective thoughts for safety and health practitioners working within the changing face of American business. Since these changes involve almost everyone, whether directly or indirectly, the rules for success are widely applicable.

3

The Evaluation and Preparation Stage: Where to Start?

My college-age son decided that he and some friends would take a trip down into a scenic National Park area during their Spring Break. The decision to go was made the night before they departed. "Where are you staying?" I asked. "I don't know. We'll just find something when we get down there," was his response. "How are you getting down there?" I asked. "We'll take Cory's car. It's big enough to take all of us," he answered. "Can his car make it?" I asked. Like the cars of most college students, his friend's car was undependable. "Oh sure, we take it around town all the time." he assured me. "The Park is almost 300 miles away. That's a little more than around town." I argued to no avail.

He ignored my rantings and ravings.

The next day, after class, he threw some things into his back pack, raided the pantry and the refrigerator, and was gone for the long weekend. That Sunday night, late, he returned. "How was your break?" I asked. "Well, it was okay," he stated as he went into his bedroom. The next evening at dinner, I asked again. "How was your trip?" The account that followed was incredible. The car had broken down on the way there. While the driver stayed with the car, the rest of them hitched a ride. After sending some help back, they hitchhiked to their destination, but there was "no room in the inn." There was no lodging anywhere. He never said where he stayed the first night, but when his friend arrived with his car (and a newly acquired repair bill), they all slept in the car or around it. The bonfires were fun, he consumed all of his supplies, and the car held together coming home. What a trip!

Unfortunately, this all-too-typical story for a college-age person is the way we traditionally approach a change effort. We make the decision and just try it. This method usually leads us to failure; or at least it yields a lot of frustration, delays, and agony.

My second-oldest son wanted to go to Disneyland during the summer. He is not as cavalier as my oldest son. He considered all the things he would have to plan to avoid worrying or wasting his money, then he came to me and asked, "Where do I start?" This is a good starting point for us.

Evaluation and Preparation

QUESTIONING IN THE BEGINNING PHASE

The change process always begins with questions. Once you decide that change is necessary, the first question is, how can you do things differently? This is the seed for the change vision. It is only natural that preparation for a change effort begins with questions. Formally, this pre-change stage is called evaluation and preparation. There are two extremes that you can take—you can ask too many questions and never get started, or you can ask too few and start too soon. Balance is the key. The purpose of this chapter is to help you evaluate your current status, the problems, the organization, and the roadblocks, and to begin a preparatory process. You may never get all of your ducks in a row; a majority of them is fine. There is no way that you can evaluate or prepare for everything.

Think of this stage as a path from evaluation to preparation. Evaluation is no more than finding out what there is to learn from the past or the present. Preparation is a positive movement, actually doing things to avoid some of the traditional problems with change and to counter traditional roadblocks.

EVALUATING WHERE YOU CAME FROM

The first lesson must come from your past. *What is your history?* Does your program come from a nonchanging environment or is it constantly in motion? Have past change efforts been successful or have they failed early? Has management supported change or resisted it? In past changes, has it been a smooth and constant transition or has it fluctuated? Your history can be positive or negative. The knowledge of what has happened before can provide information for any change you are anticipating. If you don't review and learn from your history, you are destined to repeat the same mistakes.

EVALUATING THE UNWRITTEN RULES

What is your culture? Every organization or group has a culture. The culture is defined as the unwritten rules that tell you how you can be successful and what you can and cannot do. These unwritten rules must be analyzed for opportunities and incompatibilities. Compare your vision of change to what is presently in place. How stable is your culture? Have the same unwritten rules existed since the dinosaurs or are they changing every day? Is there a match somewhere in the organization? Are there rules that run counter to your change vision? What kind of change must occur in your culture to support your change?

Evaluating the culture provides a realistic look at how your change will fit and what roadblocks your culture will throw in your path. Because culture is unwritten, it generally cannot be changed quickly, this is very different from a change in a procedure or a requirement to wear a different type of personal protective equipment.

It is important to note that within each organization's culture, it is natural for subcultures to exist; you may find a mismatch in the greater culture and opportunity in one of the subcultures.

EVALUATING YOUR CURRENT SITUATION

Too often we begin the change process without really knowing where we are and how we do things. An important point in our evaluation process is to bring two questions to the surface: *What is the baseline?* and, *What are your processes?*

These basic questions concern how different the change will be from what you are doing presently. Is your present practice top-down? Are there checks and balances, authorizations, etc., that are inefficient? Where is your time presently being wasted? Is the present situation reactionary and ineffective? What do you presently do that will be different, not done, or done by someone else in the change? What are the present staff requirements? How will these be different? What are the costs now and what will they be after the change? What advantages do you see in your present set-up? Does the change significantly improve things for you, your program, and the organization?

You must take a very hard look at these questions. When you begin seeking sponsors for a change, they will ask the same questions and you must be ready to answer quickly.

DEVELOPING A MISSION STATEMENT

Your organization or business may already have a mission statement. It tells why your organization or your function exists and how you plan on being successful. It is common in companies and organizations. Mission statements tend to be wordy and confusing. The important point for us, however, is that most programs like safety and health do not have a mission statement. You may think that a statement is unnecessary, but it is critical to a change effort. *Why does your function exist?* It is easy to respond to this with a quick answer, such as, "to contain costs of injuries or illnesses" or "to keep workers safe and healthy." But such answers are not totally correct. By answering too quickly, you may misdirect your change or not change enough. Many advantages will be realized if you take the time in your evaluation stage to analyze your function. One of the most concise mission statements I've seen for a safety and health program is as follows:

> The EHS team is committed to providing protection and prevention services by elimination or control of hazards to control human and financial costs.

As you can see, there is a certain art to writing mission statements.

You will also need to look carefully at what you envision your function to be after the change. Is it different? If so, how is it different? What impacts will that change in function have on your program, staffing, budget, and organization? These are necessary questions that you must evaluate honestly if you are to anticipate the impacts of change.

CREATING A CLEAR VISION

With your present and past in clear perspective, you need to complete the description of your vision. *Where do you want to go?* How will the processes or program be different? How will your program be different? How will responsibilities and services be different? How will the cost and staffing be different? How will management's and workers' perceptions of the service you provide be different? What will be the impact on worker safety and health? Concentrating on these details of the vision helps you anticipate problems and resistance. Have a clear vision so that it can be described in detail to others and enables you to see the change as it occurs.

SETTING THE PRELIMINARY TARGET

This stage moves from evaluation and begins preparation. *When do you want to get there?* Is this a critical change that must be in place quickly or does it require significant cultural changes that will take time? Do you want to be there next year or in three years? The target depends on the magnitude of change, the number of people and programs impacted by the change, and the amount of knowledge or skills that must be developed to implement or support the change. The preliminary target must be realistic. A target set too aggressively only invites frustration and may derail the effort because sponsors don't see change when they are told it should come about. A target set in the too-distant future may never gain momentum.

CAPTURING THE EXISTING WISDOM

Change is always preceded by other changes. Knowing this, it is important to ask, *Who can you learn from?* In school, we learn from teachers. In business, we often learn from mentors. A mentor is someone who knows something or does something well and is willing to teach you. This question seeks to identify those key people in your organization that know something of value to your change effort, who have skills, or who have championed change efforts. They are usually not at the top of the organization. If their change was successful and if it had significant impact on the organization, they may be above you on the ladder.

But don't forget to identify those who have piloted unsuccessful change and may be near the bottom of the organization as well. They too have important things to teach you.

ASSEMBLING THE CHANGE TEAM

Even Lee Iaccoca didn't change Chrysler by himself. He led the effort, but he assembled a team that worked the magic. *Who are the critical players?* This does not mean who must authorize or sponsor the change. That question comes in the second pre-change stage. This question asks who is critical to planning and guiding the effort—the change leadership team. Great advantage can be gained from a diverse change leadership team. This diversity includes levels in the organization—management, professionals, workers, etc. It also includes diversity in disciplines and specialties—engineering, training, production, quality, etc. Diversity in your team accomplishes two significant advantages: It increases perspective in the planning and implementation processes and it spreads communication pathways throughout the organization.

Once you identify your change leadership team, they must be brought together and sold your vision. If you can't sell one of your selected team members on your vision, choose someone else for your team. Your change leadership team must be committed to the vision and the benefit it will create. From this step on, the change leadership team becomes the discussion group, the planning group, the implementation group. It is from this team that the vision creates synergy and develops. This is a critical and important part of the change process.

BUILDING A SUCCESSFUL TEAM

Successful teams are not instantly created simply because the members are brought together. A critical question must be answered before you can create a successful team. *What skills do you need?* The skills may be simple or complex. They may be communication skills. They may be skills for working effectively as a team. They may be safety and health skills. Whatever skills the team or team members feel they need to be effective at planning and leading a change effort must be identified and found. Don't limit your opportunities for building skills. A formalized training class or a conference is not the only option. A skill building tool can be a book each team member gets, reads, and then discusses together. It may be inviting someone in the organization to a team meeting to share experiences. It may be identifying a team mentor, administrator, or advisor. Building a team is a slow and continual growth process. In fact, some skills can only be attained over time and some skills won't be identified until they are needed later in the change effort. Training and skill development, will be a constant part of your team effort.

LOOKING AT PEOPLE AND PROCESSES

The change leadership team must look internally at your program and staff. *What will be the impact inside your program?* It is always more friendly to look inside first. When you are considering change, what is known and familiar is far easier to analyze than what is outside and unknown. Evaluating the impact inside your program is also much more accurate, and accuracy, like honesty, is an important part of this evaluation and preparation stage. What does your program look like now? What are the processes and procedures you follow? If you don't know them or if they aren't defined or standardized, this stage is a difficult time to try to identify them. What are the present staffing and assigned responsibilities? What systems exist safely inside your program? What turf is at risk? What will your program look like after change? What will your processes and procedures look like? It is more important that you be complete in your analysis of the existing staff and processes than in your assessment of the future. But a best analytical guess is important. What will your staffing needs be? How will your systems and turf be different?

If when you compare today with tomorrow nothing will be different, don't continue your change effort. Instead, try to be a little more honest in your analysis. It is important that your team gets a close look at what will occur in your program before you look outside. If you don't see any differences in your program, you can't expect to see any differences outside.

LOOKING FOR CONFLICTS WITH TRADITION

Looking outside your program into the organization for differences between the present and expected future is also important. The question, *What will be the impact outside your program?* addresses these problems of influence. When you look at the impacts to your program, you generally have influence over those changes. Outside your program, you lose that influence. Whether those desired results happen or not, to a large extent, depends on how well you can identify where the conflicts to change will occur and on how well you plan for them. Looking outside of your program is a two-fold process. Look at the differences the change will have outside your program, then look at what stands in the way within the organization.

IDENTIFYING THE THOUGHT PROCESS

This question logically follows the previous one. *What paradigms and culture stand in your way?* Paradigms are those beliefs that define what is successful and what is not. Culture is the unwritten rules of the organization and your program. If your change vision is different from the traditional vision, there will be conflicts with existing paradigms and culture. This question encourages you to look at these roadblocks as a team.

Change is a lesson in reality. Reality is often times harsh, but it must be dealt with directly if change is to be successfully implemented. Identifying the paradigms and cultural problems that exist in the organization, in your program and staff, with your boss, or with you, are important to make visible. The sooner problems with paradigms or culture are made visible to your change leadership team, the quicker the thought-process can start on how to avoid them or successfully conquer them.

OBSERVING THE RESISTANCE: UP PERISCOPE

This is the potential enemy. Using all the information that we have acquired from answers to our other questions, we should have a very clear picture of where we are, where we want to go and how different these two places might be. Recognize that "different" is rarely met with open arms and is usually rejected. *What resistance can you expect?* You need to formally identify sources of resistance will help your change leadership team plan for effectively countering the resistance before it grows or derails your change effort.

THINKING CRITICALLY

Is the path you've chosen correct? Even after acquiring all of this information, your path may receive adjustment. The initial change path was created in a near vacuum, from a "what could be" vantage. It may be too ambitious, idealistic, or restrictive. It may encounter too much resistance or run counter to past efforts or your culture. Your path must be tuned so that it can be effective *and* successful. It is no disgrace to totally rewrite your plan in this first stage. If it is appropriate, it's critical. *What modifications are necessary to ensure success?* A journey is undertaken with the end result in mind. The purpose of change is to make things better. Success, therefore, is the difference between that better end result and spinning your wheels. It could also be the difference in upper management's perception of your abilities and, thereby, of your long-term worth to the organization.

PULLING IT ALL TOGETHER

The evaluation and preparation stage is critical to beginning a successful change effort. It is not a complex task, but requires attention to detail. Beginning with an evaluation of the past, the questioning process moves toward identification of a change leadership team. From that point on the process is team-based and preparation for change begins.

Evaluation and Preparation Questions

- ✔ What is your history?
- ✔ What is your culture?
- ✔ What are your baselines and processes?
- ✔ Why does your function exist?
- ✔ Where do you want to go?
- ✔ When do you want to get there?
- ✔ Who can you learn from?
- ✔ Who are the critical players?
- ✔ What skills do you need?
- ✔ What will be the impact inside your program?
- ✔ What will be the impact outside your program?
- ✔ What paradigms and culture stand in your way?
- ✔ What resistance can you expect?
- ✔ Is the path you've chosen correct?
- ✔ What modifications are necessary to ensure success?

STARTING A CHANGE TO EFFICIENCY IN A SAFETY AND HEALTH PROGRAM

In American organizations, I know of no other change for which the evaluation and preparation stage is more important than for total quality. This is because of large disparity between traditional American management/labor ideals and total quality. They aren't a little different. They are opposites. A change of this magnitude basically requires an organization to throw everything out and start anew. This requires a tremendous amount of faith on the part of those championing and sponsoring the change effort because their effort will be long and difficult. Faith is built from extensive communication, planning, and a healthy amount of salesmanship. These are products of a well-done evaluation and preparation process.

Many examples exist of change efforts that were thwarted because of inadequate evaluation and preparation. The automobile industry alone provides excellent case studies of inadequate evaluation of consumer needs and expectations and poor market analysis. Cars such as the Tucker, the Edsel, the Henry J, and the Metropolitan are all excellent examples of inadequate evaluation and preparation. Examples of excellence in evaluation and preparation include the Mustang, the original Taurus, and the Jeep Cherokee, to name only a few. You must do an excellent job in evaluating the past and present and preparing yourself and your plan for what you discover.

A change effort to total quality concepts cannot skip the evaluation and preparation stage. But of all the questions in this stage, none are more important than the first four: *What is your history? What is your culture? What are your baseline and processes? Why does your function exist?* From the answers to these first four questions comes the refined definition of the vision and everything beyond.

LESSONS FROM CHANGE PIONEERS

Many lessons about change can be learned from the pioneer of the American West. There were two important roles during the movement West. As Joel Barker differentiated in his video-training program

"Paradigm Pioneers," there were pioneers and there were settlers. The pioneers were those who went first, into the unknown territory to discover the bounty and take the risks. Their job was to evaluate this change movement, the movement into the West. They had, at best, only crude maps. Preliminary information (preparation for what they would encounter along the way) was woefully lacking. This is why such a high percentage of pioneering attempts failed and very high fatality rates occurred among the pioneers. It is also the reason why those who successfully pioneered the West were tough characters and gained extensive reputations from their efforts.

The settlers followed the pioneers into the West. Their success rates were significantly higher than the pioneers and their fatality rates were much lower. Why? Because the pioneers provided much information to make the settlers' efforts easier and more successful. The settlers knew not only what bounty was out there but where it was located. This information was totally from the pioneers' accounts. The pioneers drew detailed maps for the settlers. The settlers' path was easier and they had a lot less risk and uncertainty in their efforts.

This example brings both good and bad news to our discussion. The good news is that many safety and health programs have successfully changed to radically different concepts such as total quality, and from these experiences, a map can be drawn. Like the pioneers, from these successful change efforts you can learn how to minimize your work, lower the risk of failure, and decrease the uncertainty of the rewards you and your program will enjoy from total quality concepts. The bad news is that your organization is probably still rather traditional. This means, even with a road map, you'll have to be prepared to do some pioneering in your own organization.

A SUCCESSFUL STRATEGY FOR RADICAL CHANGE

From the questioning approach to evaluation and preparation, many paths can be chosen. Which are successful and which should you ignore? What path saves energy and frustration? Which problems should you accept and which should you prepare to do battle over? Who should you include in the effort? Where is the map into the frontier? You can learn much from the success and failures of others.

Most safety and health programs that consider changing to concepts such as total quality were not parts of an already implemented and successful organizational change effort. In total quality applications, what had been tried on an organizational basis was not true total quality, or it was tried in a half-hearted fashion and failed. A friend of mine calls it TQF for total quality façade. Your efforts may very well be the first to move in this change direction for a safety and health program. You will also be judged as very different by the organization as a whole. This does not mean that a successful strategy cannot be created. It only means that this is a common starting point. Therefore, the strategy will be presented toward that need.

- ✔ What is your history?
- ✔ What is your culture?
- ✔ What are your baselines and processes?
- ✔ Why does your function exist?

Most changes that have been successful in your organization have probably been top-down, autocratic-origin changes. Successful changes in American organizations have often been based on economic need, such as downsizing (or right-sizing, or the vogue—reengineering). Today's emerging concepts such as total quality, are very difficult. History begets a traditional, unchanging organizational culture, and the success of your change effort depends on defeating this enemy of tradition. Your preparation for a change must include a defensive plan or avoidance strategies because, most likely, you can do little or nothing to engage and defeat the enemy on an organizational level. If, however, you are successful in implementing a change such as total quality into your program, many examples exist in which that success has had a considerable but slow impact on changing traditional organizational cultures. It's like a drop of red paint in a bucket of white. Your success encourages other programs to try changes such as total quality (the more the total color in the bucket) and the organization's culture changes.

Traditionally, safety and health programs would fall into the turf-based and cumbersome categories, respectively. Existence of most safety and health programs is predicated with compliance-oriented and cost-containment justifications in authorization-required processes. This is a traditional baseline and is a turf-based strategy. Recognizing this structure and knowing the complexities of your safety and health processes gives you valuable insight into the effects and benefits a change to total quality will have. Generally, these will come in the form of more line or team involvement and a less authorization-based safety and health program.

It is also important for you to determine your mission statement. But think beyond what you're changing from. The mission really shouldn't change that much, even though the actual practice of the service definitely will.

- ✔ Where do you want to go?
- ✔ When do you want to get there?

Your change goal needs to be realistic. It is dangerous to say, for instance, "We want to ultimately convert the entire organization to total quality." Keep in mind a simple rule of good goal setting. Attainment needs to be under your control or influence. This is why setting a total quality goal in a non-total quality organization needs to focus on your program and staff, at least to begin with.

Set a reasonable time to accomplish the vision. If this is your first major program change effort, give yourself extra time.

- ✔ Who can you learn from?
- ✔ Who are the critical players?
- ✔ What skills do you or they need?

You must have help with significantly different change effort. There are others in your organization who know valuable information or have skills that will help you. They may have sponsored or championed some other service-type change. They may have great sales skills. They may be keen in organizing a task. The secret is to find them and learn from them.

You will also be assembling your change leadership team. This group will guide your entire change effort. Depending on your program's size it may include you and all of your staff. It should also include key members from outside your program who can provide help or resources. A good example would be a Purchasing Agent or a Hazard Communication Program compliance. Hourly or union members can be helpful if your vision includes changes in participation, ownership, or empowerment issues.

Building your team means bringing up the team's basic skill level. Consider improving your team's change leadership skills, communication skills, knowledge of the organization, etc.

✔ What will be the impact inside your program?
✔ What will be the impact outside your program?

Even after you have a realistic idea of what the change will mean inside your program, no one who is involved in leading the change effort, including your program staff will know what the true ramifications of the change are. Your knowledge of the impact of change in your program must be factual to the best of your knowledge and that of the change leadership team.

No safety and health program stands alone inside an organization. If basic changes are very different from the way you have done your services and from the way your internal and external customers interact with or get services from your program, you need to identify those differences. It is critical to the planning process to honestly communicate when discussing the changes with sponsors.

> ✔ What paradigms and culture stand in your way?
> ✔ What resistance can you expect?

Initially, you must identify obstacles in your path toward change. As discussed, traditional organizational culture is a roadblock. Other paradigms that usually get in the way of radical changes are traditional management and labor paradigms. These include the "Me Management-You Labor" participation-blocking paradigm, the "You Can't Tell Labor Everything" communication-blocking paradigm, "Good Managers Are Autocratic Managers" team-hindering and empowerment-restricting paradigm, just to name some common examples. In your list, also include the "Turf-Defined Power Base" paradigm that is common among safety and health professionals.

Resistance always accompanies large change. However, the extent of your change vision is critical to what resistance you can expect inside and outside your program. If the change vision is widely different from what customers expect, or attacks acquired turf, expect significant resistance.

> ✔ Is the path you've chosen correct?
> ✔ What modifications are necessary to ensure success?

If your list of paradigms and sources of resistance is formidable, it's okay to rethink the extent of your vision. It is dangerous to bite off more than you can chew. If, however, you have difficulty finding any paradigms or sources of resistance, you may have chosen too easy of a vision and one that will not require a team effort. The change leadership team should assess if they have the right makeup to be successful.

HINTS TO HELP YOU SUCCEED

A radical change—such as one to total quality—is very different from traditional management. Because of this, the evaluation and preparation stage is extremely important. But, there are some hints, if heeded, that can make your change effort more successful.

✎ Tuck your boss in first

The most important person to keep informed is your boss. Your boss doesn't have to be a member of your change leadership team; more often than not, it is best if he or she isn't. Your boss doesn't even have to be convinced that your change vision is the proper direction or that it will work. He or she will usually give you a partial go-ahead and wait to see the results. Ignorance on your boss's part can and will bring a quick stop to any change effort. Your good results make your immediate boss look good to his or her boss. Your boss can be a supporter or a roadblock to your efforts.

✎ Begin small and internal

Don't take on too much at once. This hint follows one of the rules of good goal setting—make sure a proposed change is under your control or influence. Remember, your boss and others around you will be watching to see what comes from your change effort. You don't want them waiting too long and you don't want them to see you struggle or fail. If you begin small, for example, in your work team or industrial hygiene or safety program, you focus on an area where you can succeed. Tomorrow, you can widen your focus.

✎ Don't make assumptions

We often make indirect assumptions. We don't want to be accused of asking a stupid question. By not asking every question we can think of, we make dangerous assumptions. Assumptions come totally from our own frame of reference (FOR), which often suffers from inadequate information. If you ask all of the questions, all potential aspects of change are made visible to your change leadership team. More often than not, someone has a different view of the answers, which is derived from their own particular FOR.

If the question isn't brought to the surface, you make assumptions about the past or current situation, and about what sources of resistance you may encounter. This causes significant flaws in the change planning process and, thereby, leads to many surprises in implementation. These surprises can be easily avoided if we don't make assumptions early in the change process.

✎ Don't be surprised by the size or strength of the opposition

Many a military general or world leader has underestimated an opponent and paid the ultimate price. Examples include Napoleon at Waterloo, Lee at Gettysburg, Hitler in Russia, and three American presidents in Vietnam. These are poignant examples of miscalculating the enemy. Anything that stands in the way or slows your change is your enemy. If you underestimate your opposition, you can't strategize adequately to win. You could then pay the ultimate price by failing in your change effort.

✎ **Expecting the worst is far better than minimizing**

It is far better to have mobilized more than enough of your forces to meet a particular foe and to be surprised how weak the opposition is than to show up with a light brigade and find 50 million barbarians. It is better to say that it wasn't as bad as you thought, rather than worse than you expected. What is the worst possible thing that could happen or be encountered? In your vacation planning, it's like saying, "What will we do if the car breaks down, we're 300 miles from the nearest service or town, and you get a snake bite?"

Visualizing the worst possible (reasonable) scenario helps us to plan for that occurrence. Such plans may include expanding your change leadership team, getting additional change sponsors or resources, increasing specific training, or including the union in your discussions. Whatever the solution to your worst possible situation, it needs to be part of your evaluation and preparation process.

✎ **Make economic comparisons important and well thought out**

You may be thinking that changing to total quality will have a large impact in your safety and health program or will ensure less frustrating or reactive dealings with management. These are, after all, well-known advantages of changing to total quality. However, don't assume that these advantages carry the same weight with upper management. They don't walk in your shoes daily and may not be aware of the need for change. Management also tends to see things, at first, from an economic viewpoint. Therefore, you need to be ready to meet their expectations when you sell your change vision, seek their sponsorship, or request resources. Management that thinks only in economic terms is a reality in traditional organizations. Be prepared to talk about economics if you want to be perceived as a visionary and capable leader of positive change.

✎ Diversity in teams transcends bias and shortsightedness

What is one of the greatest advantages of using a team-based approach to change leadership? It allows you to gain perspective and diversity in your input. Too often we see things only from our FOR. Diversity in your change leadership team provides many different FORs. These differences strengthen your evaluation and preparation stage answers by exponential proportions. Areas where you can add to the diversity of your change leadership team include having members from different service areas such as engineering or purchasing, customer groups like production or maintenance, and different organizational levels including hourly employees and management.

✎ Treat this change effort as if it is your only chance to succeed

How stable is your position or program? How much micromanagement do you get to "help" you with your safety or industrial hygiene job? What is the level of your reputation in the organization? These are frequently unknown quantities. That's why every opportunity to succeed is important. Besides, a change like one to total quality can provide so many benefits to you and your program that you really cannot afford to waste the effort. In reality, you may not be authorized to try again.

With a change vision like total quality being so much different than traditional American management, failure may doom any second effort in your organization. This hint says simply, "If you are going to think about a change like total quality, you had better take the effort seriously." It may have more gravity than you might think.

✎ Be a cheerleader

Everyone loves a cheerleader. There are overwhelming positive perceptions within our society that confirm this. Cheerleaders are energetic, exciting, and their attitudes are contagious. Remember how you felt when you attended a college or high school sports activity? This is really a comment about how to be a good change champion. A champion of change is the person or persons who sell a change vision and keep the change effort going. Obviously, you can be a quiet champion and work behind the scenes. In a radical change like one to total quality, and being in a service-oriented program like health and safety, you had better be a champion who is also a cheerleader. A drastic change like total quality requires a more highly visible, positive, and energetic approach to the effort.

✎ Never downplay communications

Change usually involves some negative impacts to the organization, people, or functions in the organization. You can't make changes positive for everyone. Recognizing that, there is an urgent need to communicate the change vision *and* the expectations or results once the change is in place. It is easy to talk about the positive results. It is not easy to detail the negative results, especially to those who will be impacted by them. Change is a time for honest, even painfully candid, communication. In fact, it is more important to talk about and emphasize the negatives than the positives. This approach prevents surprises, and high levels of resistance due to the surprises, and helps those impacted deal more openly with negative results. Keeping things out in the open gives you credibility (not popularity) and helps to prevent covert communication.

The evaluation and preparation stage is the most important stage in a change to total quality. It involves answering tough questions, getting a lot of people involved, getting the blessing of your boss, and communicating honestly. Considering the benefits that total quality can bring to your safety and health program and to you as a service provider, these difficult changes are well worth the effort.

4

The Solicitation and Creation of an Environment for Change Stage: Preparing the Field of Play

In recent years, quite a few NFL teams have moved to other cities. When the Cleveland Browns left for Baltimore, the negotiations became highly complex and the settlement set new precedents of how much can be negotiated. Generally, there are strong reasons for an NFL franchise to want a new stadium and city. For them, the change makes good business sense and is worth all the trouble that moving a franchise causes. In reality, the new stadium, fans, community support, and financial arrangements and contracts are the necessary components that make up the elements of a new field of play.

Making an analogy to our change process, the negotiating process that makes the move to the new field for the NFL team is the solicitation process that prepares the way. The stage of the change process that includes solicitation and creation of an environment for change gains a lot of clarity.

Although complex written contracts are not usually an accompaniment with an organizational or program change, many verbal contracts are made during this stage. As in moving an NFL team, there is considerable planning and salesmanship involved in solicitation and creation of an environment for change.

Solicitation and Creation of a Change Environment

PLANNING FOR SUCCESS

The purpose of this stage is simple and two-fold. First, you need to acquire the appropriate level and line of sponsorship for the change. Sponsorship is a larger concept than mere approval. Approval is a head nod. Sponsorship is an active participation role and involves the provision of necessary resources. A sponsor is a key player in the change process. A sponsor also often has to "go to the well" in support for a change process, especially when resources are needed, strong resistance is met, or resistance surfaces from inside management.

The second purpose of this stage is to begin the process of preparing for the change. Specifically, this stage examines the personnel and other resources that will be needed in your change effort and begins a continual process of receiving and evaluating feedback.

Because of the importance of using active sponsors and avoiding or minimizing resistance and roadblocks to change, this stage is crucial to a successful effort. This stage cannot be avoided, even for changes inside a program over which you have total day-to-day control. Even inside changes cause ripples outside of your area of influence which will bring about resistance. If you don't have sponsorship, resistance from an influential portion of (or person within) the organization can quickly reverse your progress.

Sponsorship of positive change can also be an effective tool if you want to show anyone above you that you are a team player or that you have promotability. Being known as a successful change leader, a good team player, and a creative thinker will never hurt anyone in an organizational streamlining or downsizing effort.

OBTAINING SPONSORSHIP AND OTHER CHAMPIONS

What do these two terms mean—champion and sponsor? For our use in implementing change, knowing the difference can be important. In change, champion and cheerleader are almost synonymous. At a sporting event, cheerleaders provide and lead the cheers that motivate the team. They get the spectators involved in the activity. They are entertaining and full of energy, and they are persistent. Even when the team is losing, they lead cheers throughout the game. Cheerleaders also tend to be popular with others because they are involved and optimistic.

Let's draw some direct comparisons to change champions. First, they provide the clear vision that will motivate the team and sell the plan. Second, they get others involved in the change activity. Third, they are entertaining and full of energy. Fourth, they are persistent. Even when the change effort is going badly, they continue to share the vision throughout the effort. In the beginning, they are not popular because they are the change personified; but, if successful, they become icons within the changed organization. Generally, in successful cases, they become very popular with upper management.

Picking the right people to sponsor a change is extremely important to your change efforts and to the chances of successful change. Pick them wrong and your efforts will be up hill, into the wind, and with a high risk of failure. Pick them right and you can coast to victory. This change step must begin with a fundamental question: Who has line responsibility over the area or areas that the change will impact? If the change is organization-wide, the answer is simple—it's the person at the top. If, however, it is a change to your safety or health program, that change has tails. Many may need to sponsor the change. Of course, the change will have ultimate impact under you and your boss' line responsibility. But how does that change impact customer expectations or how the service will be provided? The definition of line responsibility in these cases can be much more complex.

Figure 4.1 gives a simple example of a line organization with the traditional line structure (Manager A through Workers) and a service-oriented staff function. Using this figure, let's consider three examples.

Example 1: Who would be the appropriate sponsor(s) for a change at the Staff Program area of the organization? Naturally, the Staff Manager would have to be an active sponsor or a champion of that change effort. Manager B would also have to sponsor the change as a result of the reporting structure. Manager B's sponsorship role would be more informal and less involved than would the Staff Manager's.

Example 2: If, however, the impact of the change effort at the Staff Program area also involved changes to expectations in the line structure, the responsibilities in the line structure, or how that staff

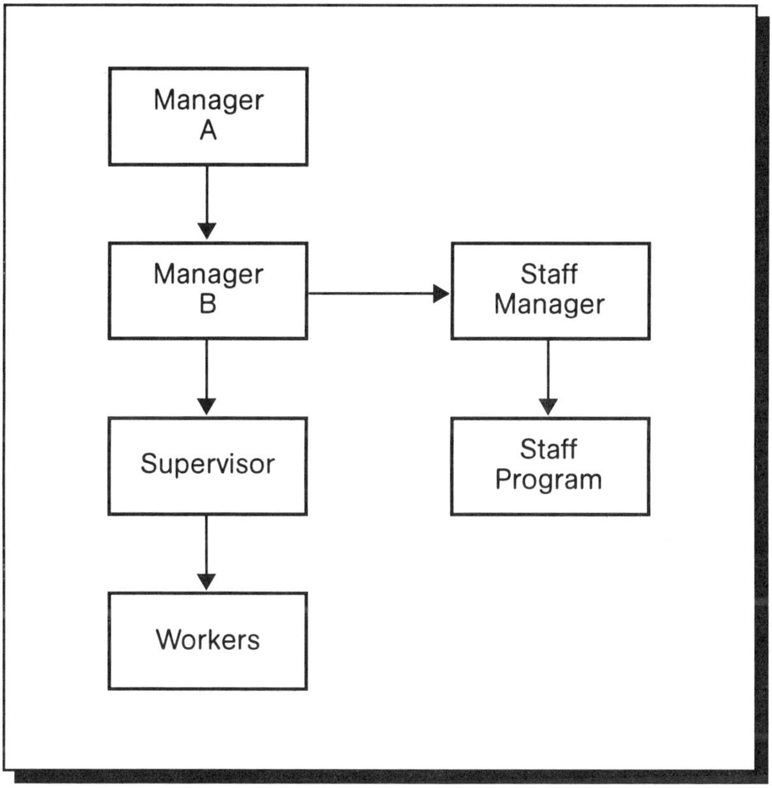

Figure 4.1. Traditional Line Responsibility Structure.

function's specialty is conveyed to the line structure, the sponsorship issues change. Not only would the Staff Manager be an active sponsor, but Manager B would have to be involved too. Additionally, Manager A would need to be a semiactive sponsor of the change so that resistance and resources could be appropriately dealt with and provided, respectively.

Example 3: Organizational changes would have to be actively sponsored by Manager A.

These examples also distinguish between direct and indirect sponsors. In Example 1, the Staff Manager would be the direct sponsor of the change, while Manager B would be an indirect sponsor. In Example 2, the Staff Manager continues as a direct sponsor but Manager B, because the changes have impact within his line of responsibility, must also be a direct sponsor of the change. Manager A would serve a split role. At times, Manager A would indirectly authorize the active change effort but at other times he or she would have to be a direct sponsor when resistance or need for resources required involvement. In Example 3, the organization-wide change, Manager A is the direct sponsor and the key to the success or failure of the change effort.

Once sponsors are identified, how do you actually obtain sponsorship? Most often there is more involved than simply asking. Logic tells us that merely having a great idea would gather sponsorship. Our experiences with new inventions and thought processes, however, quickly convince us that this is not the case. For example, Chester Carleson's new photographic technique eventually became electrostatic copying or the Xerox method. Time after time he presented his idea to heads of photographic companies and it was rejected. W. Edwards Demming had ideas for change involving statistical process control. He was sent to Japan. If your idea deviates significantly from what is traditional or is widely different from prominent paradigms, beware. Your change vision, sometimes on faith, other times at personal risk, will have to be sold into a "wait and see" format to obtain sponsorship. In a less than enthusiastic sponsorship arrangement, your preparation and implementation plan had best be excellent because sponsors may not readily offer resources.

The importance of appropriately selected sponsors cannot be overlooked when planning for change. Appropriate and supportive sponsorship is the catalyst that drives or retards the effort's chemistry.

IDENTIFYING AND OBTAINING NECESSARY RESOURCES

Any successful change effort requires resources, which are generally controlled by your sponsors. You should consider this when choosing the appropriate change sponsors. If, for example, internal organizational communications is controlled by a staff area outside of your direct or indirect sponsorship analysis, you may want to include the head of the communication service as one of the necessary sponsors. When you are seeking sponsorship for your idea for change, you are attempting to build a comfort level for the change vision in those individuals. If you are successful, they will generally sponsor your change if it doesn't violate any formidable elements of organizational culture or their paradigms. It will also be your responsibility to keep their comfort level high. This requires continual attention to and communication with your sponsors.

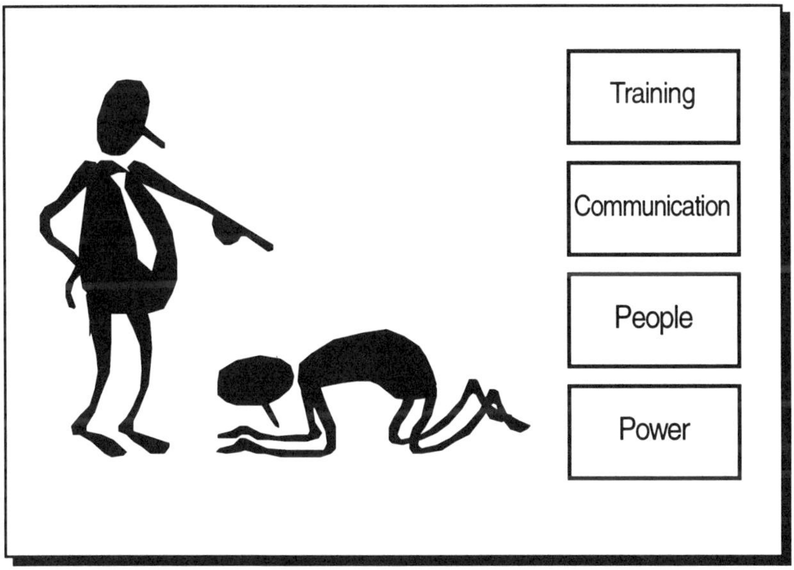

Four critical resource areas exist in a change effort: training, communication, people, and power. Because change is different it requires that new knowledge and skills be built within the planning and implementing stages of the change. Without divine knowledge or osmosis, there is only one way of obtaining knowledge and skills—*training*. The definition of training is not limited to the traditional classroom or conference mode. Training has many modes and faces. It can include using a key organizational player as a mentor; learning from a successful or failed change effort; reading a book on change, planning, or team dynamics; training via a video program; talking to someone on the phone. The point of training is to learn. You should direct your attention to learning about a particular area of need and choosing the most appropriate mode for that training.

Some training is free, while other training is very expensive. If your change sponsors need to authorize an expenditure for training, and if they are unsure of the plan, it can derail your change effort early by removing training opportunities and you may be labeled as a spender, not a saver.

Communication is a critical resource in change. Because information is traditionally considered a major source of management power, it is generally tightly controlled in any traditional organization. At the least, overt communication is controlled. The communication resource must be conveyed by sponsoring management. One of the most destructive elements of communication in a change process is in the covert arena. Once information reaches the grapevine, you can't control the accuracy, the repetition, or the timing. Covert communication is seldom supportive and rarely positive. Overt communication is very important in the change process. It helps everyone know what is going on, what the change vision is, why that change vision is good, what progress is being made, etc. Communication is critical to a change effort and overt communication pathways in any organization are controlled by management. Only through sponsorship in your change vision can you use communication to help you succeed.

The smaller your change effort and impact, the fewer of the third resource, *people*, you will need. If you are implementing a change in a small group only, you may be able to handle the change yourself. If your change impacts organizational functions, you will need a number of people working with you. These people are called change agents, and they interface with the area of change and implement it. For example, if you are changing the way safety inspections are done by involving hourly workers, maintenance, and supervisor inspection teams, your change agents have to be those who are directly impacted by those changes. Obviously, production and maintenance supervisors and managers will need to be enlisted as change agents. People resources can only be acquired if line management directly sponsors the change.

Sometimes our parents' approach was very applicable to change efforts. "Why?" we asked. "Because!" they replied. They had a position of *power*. Sometimes power is the only way to force a beneficial change into being. Shigeo Shingo, the father of Japanese management, probably said it best. "Some people can't change." A plant manager colleague of mine says it differently, "Sometimes you need to give them a chance to sell suits at J.C. Penny's." The operative phrase here is "beneficial change."

A friend of mine had to kick his son out of his house after a considerable amount of money turned up missing. It wasn't the first occurrence. No money had disappeared between the last time his son had left home to be on his own and his return. This was a difficult decision for my friend, but sometimes there is only one answer that is best for everyone concerned. Luckily, it turned out okay for my friend and his son. It could have ended very differently.

Keep in mind that power should be used as a last resort. First, make *every* attempt to make change happen without using your power button. When all else fails (not when we become impatient), then push the button.

Power is controlled by management. As the power base continues to erode in organized labor (as evident at Caterpillar and other companies), we can't forget that. Beneficial change that confronts an irrational, unreasonable, and unmoving force must be assisted by power.

ENTRENCHING YOUR MISSION IN THE ORGANIZATION

We often make dangerous assumptions. In change, especially if it is disruptive or differs a great deal from what is traditional, some assumptions can be lethal. This is commonly the case for our mission in safety and health. If we have completed step one of the change process, we know our function and how we are going to make that function valuable to the organization. We derived our mission statement. At this point in the change process, we cannot assume that our mission is common knowledge within the organization, even to our immediate boss. When we are actively planning and selecting sponsorship for our change effort, it is the perfect time to also stress why our function exists and how it is important to the success of the organization. This relationship to organizational success is called adding value in business.

Entrenching your mission in the organization is a lot like preparing a garden. A change vision (seed) left naked on the surface cannot take root and grow, and thereby, has little chance for survival and success. When soil is properly prepared for the seed, the crop is more likely to be successful.

Whether he or she is a direct or indirect sponsor of your change effort, your boss should be told why your function exists and why it is important to the success of the organization. This information is a critical foundation upon which you will lay your vision of change. Your sponsors must connect your function to the organization's needs in order to be successful. From this sponsor foundation, your mission must be passed throughout the organization. For example, it is important for the workers to know how you perceive that your function will be accomplished after the change. This is also important to front-line management, especially if there are to be changes in the way your service is provided to the line or the way your function is perceived.

This is the transition step between soliciting and creating the environment for change. Obtaining the dedicated and enlightened support for your purpose and approach from your boss and the appropriate change sponsors is the spring-board to the next step which begins the process of creating that change environment.

SHARING THE CHANGE VISION

Sharing the change vision is like hosting a grand opening. You begin by selling your change vision to your identified sponsors. Nine important questions must be answered, as seen on page 96.

Generally, sponsors need as much information as you can give them in as concise a package as you can place it. There are also a few simple but often overlooked points that you need to be keenly aware of. First, you'll get no benefit from being overly detailed or lengthy in your presentation. Management tends to get impatient with too much detail (there is a delicate line between too little and too much) and an Executive Summary format is preferred. Be quick and to the point.

What Your Sponsors Will Want to Know!

1. Why can't things stay the same?
2. Precisely, what do you want to change?
3. Exactly how will it look or work once it is changed?
4. What other options did you look at?
5. What are the advantages/benefits of this change?
6. What are the disadvantages of this change?
7. What general approach to implementing the change do you intend to take?
8. What resources do you need to be successful?
9. When do you expect to finish your effort?

Second, your sponsors need the truth even if it's murky, arguable, or negative. If they don't hear the truth from you, or if they get a "rose-colored view" from you, you are hindering your own effort! Your integrity is critical in a change effort. Third, don't share your vision until you are ready and have all the answers (or at least as many as you can at that point). Beginning the effort prematurely only labels you as naive. Fifth, as Paul Hornung said in his 1970s TV ads, "Practice...practice...practice." This may be your one shot at winning sponsorship for your idea. You need to be ready for the challenge. Sponsor information follows a *Why, What, How,* and *When* format. It is a clean, straightforward approach to providing all the information needed to gain support for your change.

Once you have sold your sponsors on your change vision, it is time to share it with others. Generally, the first priorities are those directly impacted by the change. Second those who are in the line structure above the point or points of impact. Third, include those in the organization that are just interested. And fourth, if you want to make any change

Never Forget!

- Keep it Short and to the Point
- Tell Them the Truth
- Get as Much Information as You Can—Before the meeting
- Be Well Practiced

movement in the organization or to get recognition for successfully implementing a change, spread the information around the organization.

What Others Will Want to Know!

1. Why can't things stay the same?
2. Precisely, what do you want to change?
3. Exactly how will it look or work once changed?
4. What are the advantages/benefits of this change?
5. What are the disadvantages of this change?
6. What general approach to implementing the change do you intend to take?
7. When do you expect to finish your effort?

Obviously, your sponsors need different (often more) information from what others require. If you give others too much information, you may very well be arguing and negotiating your change vision with those who are too invested in tradition to be objective. Even so, this information list begins with the most important piece of information, *Why. What, How,* and *When* are also provided. The details of what other options you investigated and what resources you need or plan to use are, of course, missing. The first missing part is not negotiable at this point in the process. You are the champion and your sponsors are backing your plan. The resources are omitted because your target audience cannot offer or appreciate the importance of these necessary resources.

COMMUNICATING THE IMPACTS

A critical part of sharing the vision involves communicating the impacts of change to all levels—both sponsors and others in the organization. Impacts fall into two areas: benefits or advantages, and costs or disadvantages. Too many times change efforts have been thwarted midstream because only half of the story was told early in the process. Usually, benefits of the change are stressed. When the disadvantages are made known, resistance hits a crescendo at all ranks, including sponsors.

Never surprise a sponsor. A predictable and rapid process occurs in sponsors who feel that they are left holding an empty bag—they quit listening, they quit believing, and they either remove their support or join the other side. Any negative aspects about your change that are remotely possible should be aired early in your communication with sponsors.

Three other rules exist when communicating impacts of change on those in the organization. First, *Perception is reality.* The only variable may be time. What people perceive is the truth is hard to change, once the perception is formed. Eventually, perceptions will change, but it may be too late for your change effort. Keep others informed before they get incorrect perceptions. Second, *Be honest.* You get innumerable opportunities to tell the truth and gain trust, but only one opportunity to lie. From there on, they won't listen. Third, *Tell them the whole story.* We are creative animals. Imagination can be your worst enemy. If you don't give enough information, others are forced to fill in between the lines.

When communicating the impacts, six basic information areas need to be communicated.

Communicating Impact

1. Benefits (Rewards)
2. Changes from tradition
3. Changing skills and knowledge
4. Aspects of participation (Roles)
5. What they will see
6. When they will see it

Communicating *benefits* is an easy and obvious area. Keep in mind, though, the frame of reference of the person receiving the information. Talking about "This will be great for our company," or "It will allow our company to increase it's market share by 30%," is fine, but the truth is we all want to know, "What's in it for *Me*?" Your message about benefits needs to be heavily focused on individual *rewards*. First, changes that have no believable or personal benefits are hard to sell. Second, don't be ambiguous. "The company will flourish and this will make it possible for most of us to keep our jobs." The response might be: "WHAT! Did he say that I'll be out of work if we change?" That wasn't the message, but it was what they heard.

Communicating the *changes from tradition* is also an important impact of change. We live by predictability. The bad news is that if you disturb someone's predictability, it is natural to resist that change. The good news is that if the person is told beforehand, the resistance may very well be less and the response to the change will definitely be different. This opens doors for your change, rather than closes them before you start. Changes from tradition include any changes in expectations—theirs, management's, and the company's. It also includes changes in practices or the way things get done, changes in responsibilities, work roles, etc.

One of our basic fears in change is "Will they need me once the change happens?" or, "Will I fit into the new way of doing things?" Communicating *changes in skills and knowledge* tells those impacted by change four things. First, you've thought this through completely; second, you are planning to keep them. People don't expect to go from one area to another without some sort of cost. If there is something they can learn or a skill they can develop, tell them. The only other possibility would be hiring new people. This message also tells them that, third, they are important to the organization and to the organization's success, and fourth, that you are willing to invest in them. They are, in fact, a valuable resource to you and your change plans.

Communicating the *aspects of participation* tells others who else is in the game. It's like a program at an athletic event. Who is leading the effort (key champion), who is or are the sponsor(s), who is involved in implementing the change (change agents), and any other key players, including the change leadership team. It is also a time for communicating the individual *roles* in the change process. Everyone has a role, whether it is monitoring change, communicating, implementing, learning, or just changing. Participation is a critical element to successful change. The more people you have working toward the change, the higher the probability of successful implementation, the fewer the roadblocks and less the resistance, and the shorter the time span that is required.

Your role is similar to that of a travel agent. You have, by this stage, given others a look at how you envision the future. You've also told them what they will need to pack (new skills and knowledge) and who will be their tour guides. Now you need to tell them about their trip. You may tend to accentuate the negative aspects, especially if you are a little unsure about the change. Talking about *what they will see* along the change path does three things. First, it tells them that you have knowledge, so you must be a leader to be followed. Second, it removes some of the unsettling unknowns that may keep others from getting involved. And third, as they see progress, it increases your credibility, their participation, and their trust, and reduces the possible sources of resistance.

SELECTING CHANGE AGENTS

A change agent's role is critical. The selection and preparation part of the process is paramount. Change agents are on the front lines of the change effort. They directly implement the change at the change target.

You can never have enough change agents. You can, however, have too few. A saying was displayed on President Ronald Reagan's desk: "There is no end to what a person can accomplish as long as they don't care who gets the credit." This emphasizes the importance of change agents. Paraphrased for a change process, the saying might read, "There is no end to what a creative change champion can accomplish as long as he/she gets a high level of participation in the change process." Too often, change agents are lost in the effort. They simply get their job done. They are very unobtrusive champions and deserve recognition for their important roles.

If change agents are so important, how can you be sure to select the right ones? Some simple rules about selecting change agents may help. Two simple organizational structures can also be used to illustrate the many important issues for change agents.

Change Agent Selection Rules

1. Look One Level Up from or Inside the Target Level
2. Must Have Line Responsibility for Targets or be Respected Peers
3. Must Have Daily Contact with Targets
4. Must Have Established Accountability Structure
5. Must have a Line to an Active Sponsor

The first rule, *look one level up or inside the target level*, can best be illustrated by Figures 4.2 and 4.3. In the first, the worker's level is the change target. The obvious change agents are the first-line supervisors or foremen. In organizational-wide changes, 99% of the time this is the change agent level. The figure provided illustrates a change that is championed by a staff function such as safety and health. The second figure shows the change roles as they might be for a staff program change. Note that the Program Manager wears two hats, as does the Program Specialist. This illustrates change by team participation.

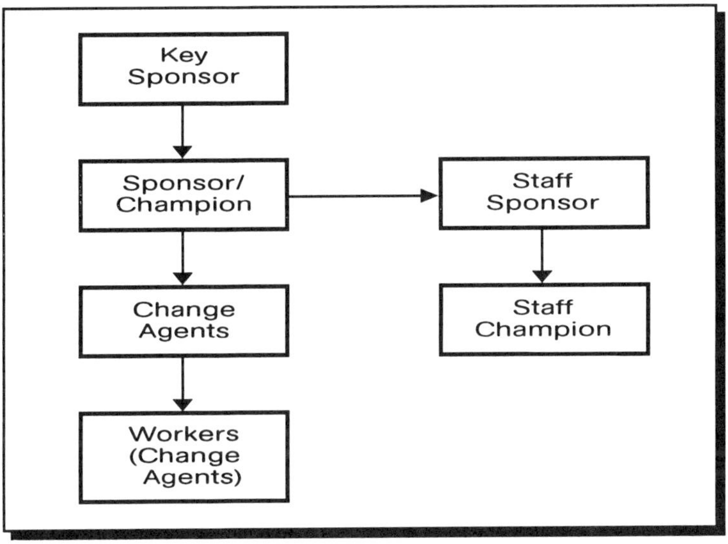

Figure 4.2. Simple Staff-to-Line Roles in Organizational Change.

The second rule, *must have line responsibility for targets or be respected peers*, describes the formal and informal ways that the change agents might interface with the targets. Both can have advantages and disadvantages. In changes that require strong influence, the line responsibility works better because of the supervisory orientation to the

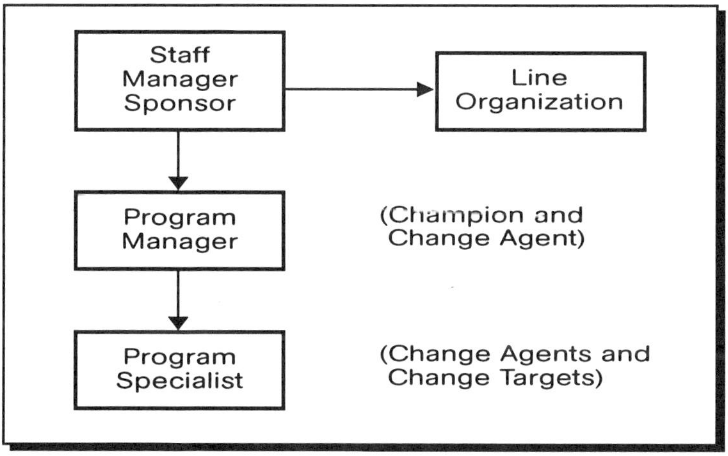

Figure 4.3. Roles in a Simple Staff Program-Only Change.

targets. In changes that get power from esprit de corps or that feed on synergy and involvement, peer participation is unbeatable. If, however, you can get support from both levels, you can get the advantages of both with few, if any, disadvantages.

Must have daily contact with targets, the third rule acknowledges that we are human. Absence does not necessarily make the heart grow fonder. In change, absence of day-to-day contact allows targets to forget and feel that they are not accountable for changed behavior.

The fourth rule, *must have established accountability structure*, means being able to couple responsibility with accomplishment. In most line organizational structures, members know what responsibilities each other has and, ideally, they are held accountable for accomplishing those responsibilities regularly. It's when responsibility and accountability are uncoupled that problems begin. When important jobs or responsibilities are not done, there is no accountability to pull it back into line. And when accomplishment is achieved, there is no fanfare.

The last rule, *must have a line to an active sponsor*, helps a great deal. It ensures that there is line involvement and commitment to the change. Without this line commitment, change agents become disenchanted easily because the change is not valued by their boss.

OBTAINING FEEDBACK

This is a time to have your antenna high and receiving. Keep in mind that it is natural to hear a lot of negative feedback. Change is always resisted, even positive change. Listen well but don't get your feelings hurt. A change vision is a deeply personal and emotional item. To be an effective salesperson, you have to really believe in your change vision and have some passion for it. This is emotional involvement. Even so, you must learn not to take negative comments personally.

Comments will probably run, depending on how controversial your change vision is, between 70 and 30 to 95 and 5, *against*. This is not bad! Others may be saying that they don't have enough information, need a clearer picture, or need information repeated. They may be saying that you should sell the benefits a little harder. They may be saying that they don't have the knowledge or skills to move forward. They may be saying that they like the change, *but* . . . . Listen carefully to what you are hearing, especially from your sponsors, other champions, and your change agents. Then, respond appropriately by giving them what they need.

Throughout the change process, feedback is an important source of information. Feedback is your best friend. If you hear what is being said, it allows you to react to it and make changes or modifications. The more feedback you get, the better you know what's going on, the easier you can respond to it, and the less of it will go underground.

DEALING DIRECTLY WITH CONCERNS AND FEARS

Most of us would rather avoid dealing with concerns and fears. Change isn't a logical thing about which you can argue the pros and cons. Keep in mind that it is difficult, if possible at all, to separate our logical existence from our emotional side. It's from our emotional side that our creativity comes and our excitement percolates. It is also the source of our concerns and fears.

In dealing directly with concerns and fears, we must keep one thing is mind—you can't battle emotionally attached aspects of life with logic. Many husbands and wives have tried that ill-fated approach with their spouses and many (if not most or all) have failed. The real pity, though, is that men haven't seemed to learn anything. But we must also realize that it is far better to deal with concerns and fears directly than to ignore them.

REEVALUATION AND PATH MODIFICATION

It is not a sign of weakness to say that you didn't get it entirely right the first time. This is not failure, this is course correction. There are three possible outcomes. You can turn back, you can go full-steam ahead, or you can change course to varying degrees. Once you've shared your vision for change, you can still modify it. Feedback will tell you if you are on the right track. Do you have or were you able to obtain the right sponsorship? Could you find other champions for your change? Do you have an effective cadre of change agents to accomplish this

task? Is the organization supporting your efforts? Is your change meeting with any cultural resistance? Does your change timetable seem correct? What have you forgotten or overlooked up to this point?

Applying feedback to questions such as these helps to determine which course is appropriate. Is the change right but the timing wrong? Is the change just wrong for what you wanted to accomplish or for this organization? Are minor or major course modifications necessary in order so that you can succeed?

This is the pause-and-reflect point of the process. It is the last step in the pre-change stages. Before we embark on the more active change-activity stages, it seems appropriate to fully reevaluate the road to this point and beyond with our present knowledge.

CHANGE TO EFFICIENCY IN A SAFETY AND HEALTH PROGRAM

This is the stage when you realize that new management concepts such as total quality are very different from traditional American management thought. Most operations up to this point in the change process have been developed in-house, in your head, or were those of your change leadership team. At most, you might have shared your change vision with a mentor or two, maybe your boss, and certainly your change leadership team. But, to a great extent, the change vision has been protected. In the solicitation and creating an environment for change stage, your change vision and general approach to implementation see their first light of day. This is when you run smack-dab into the undeniable conclusion that these concepts push a lot of traditional American management paradigms to the limit.

This is not to say that only those of strong will and firm resolve should continue past this point. It is just a fair warning. It is also a call to do you homework; prepare and practice very well. This advice is not, "Turn back, Sailor! There's a reef dead ahead;" rather, it is, "Make sure of your course and, if necessary, be prepared to do some swimming around sharks."

5

THE PLANNING AND IMPLEMENTATION STAGE: GETTING MOVING

My mother-in-law and I are opposites when it comes to vacations. I couldn't stand to take a trip the way she does and I'm sure that my method equally unsettles her. Neither method is wrong and both seem to work equally well for vacations. Her way is to plan absolutely everything. She plans when to get up, where to go, how to get there, when to return, when to eat, everything—*before* she ever leaves for vacation. I find that method a bit tedious. The major reason I go on vacation is to relax. She disagrees, though. She feels it is more important that you make efficient use of your limited time and do as much as possible. I don't care if I miss some things if I'm enjoying others or just feel like sleeping in, for example. I'm sure that she feels my way is a little too open-ended. Oh well, different strokes for different folks, right?

There is a logical reason for my approach to vacations. In my working life and in my private life, I tend to plan, execute, evaluate, adjust, plan, execute, evaluate, adjust, etc.; it's a natural cycle for me. If it isn't on the schedule, it goes onto a priority list. If it measures up to the other priorities, it gets some time; otherwise, no time is allotted. Vacations offer me relief from cycle and priority lists.

What can we learn from these two vacation-taking methods in our search for mastering successful change implementation? Suppose we applied the two methods of making a change. My mother-in-laws method would not allow a response to any unforeseen occurrences and has no mechanism for changing direction. In short, it is a *strong* planning focus.

My way wouldn't work much better because it would be so reactionary to the unforeseen occurrences that nothing positive would be accomplished. In other words, we would spend all our time reacting and none making progress on change. My way is a *weak* planning focus. This illustrates that one must plan a change carefully and in detail, but there has to be built-in mechanisms to sense unforeseen happenings early and tremendous flexibility so that the new occurrences can be dealt with.

Planning and Implementation

THE INTERRELATED NATURE OF PLANNING AND IMPLEMENTATION

If there is a theme to this stage it is the interrelated nature of planning and doing. They are not stand-alone issues. Planning begins first, overlays implementation, continually absorbs and responds to feedback from implementation, and then continues past the point of successful implementation—the finish line. They are, in fact, so closely related and dependent upon each other that they are part of the same character, like brain and body.

Let's use a simple analogy to illustrate this further. Consider a baseball player preparing for his or her turn at bat. Preparation for the turn at bat begins in the dugout. The player analyzes the pitcher's movements, the travel of the ball, the timing of the delivery, the pitcher's selection of pitches, the placement of the pitches over the plate, etc. The player's mind is deeply engrossed in planning. Mentally, the player times the pitches, sees his or her swing, visualizes the connection with the ball, watches it head toward the fence. The player even imagines the pitcher's head following the arc of the ball and the fielder slowly jogging toward the fence helplessly. Selecting his or her bat, the player begins to move from planning to implementation as stretching is done and practice swings with a weighted bat are accomplished. The planning continues as the player kneels on the on-deck circle. Again he or she visualizes the swing and watches the ball placement as it crosses the plate. Planning moves to practicing implementation as timing is synchronized while pitches are delivered to the batter. Finally, it is time for the real implementation. Now the player is the batter. The finishing touches are added to the implementation preparedness as the player-turned-batter digs into the batter's box. Gentle practice swings are made. Eye contact is made with the pitcher. Another practice swing made. The batter sets into a ready-to-implement stance with the bat cocked high above the shoulder. The pitcher accepts the signal, sets, winds-up and makes the delivery. Preparation and implementation are now synchronous. As the batter watches the delivery, speed, and travel of the ball, visual feedback is being processed by his or her brain and the planning is modified from those inputs. The feel of the hands gripping the bat and the beginning of the batter's body movements into the swing are monitored by the brain. Micro adjustments are made in the swing, the body mechanics (the implementation). The eyes watch the ball as it continues toward the catcher. As the swing continues, its height and velocity are continually modified. Implementation stops the instant the swinging bat meets the ball. Crack! The planning part continues as the batter watches the result of the implementation. Planning begins first, coexists interconnected and dependent upon implementation, and continues beyond.

Successful planning for change needs to be inserted into any ongoing program. Detailed and long-term planning is not common in the practice of safety and health. This may very well be one of the changes you might be considering. In any case, planning for change is independently determined with a specific focus on the intended change, and then inserted into the ongoing program plan. Because of this, there are some necessary components you will need in your change plan. Some helpful hints will increase the effectiveness of your planning process.

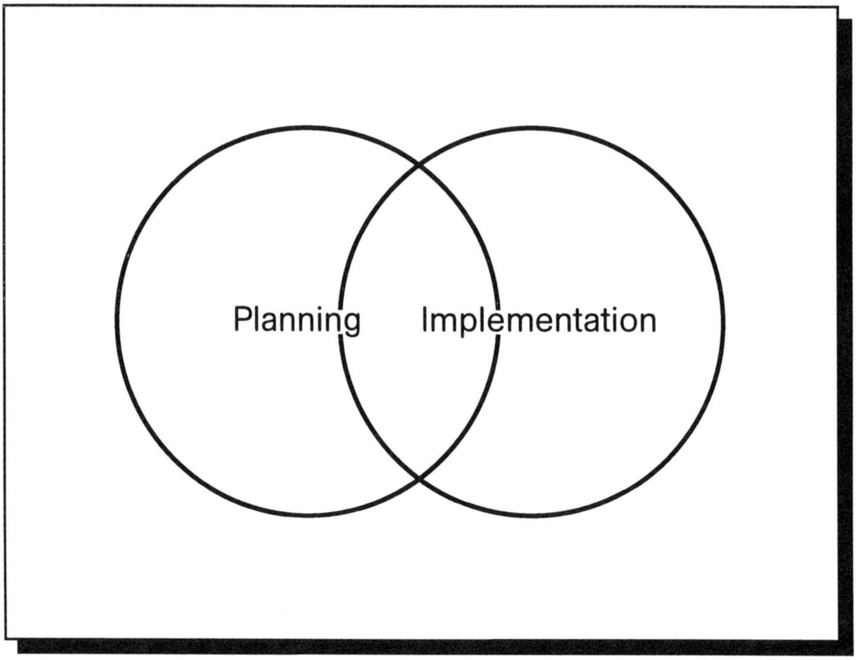

Figure 5.1. The Integration of Planning and Implementation.

PARTICIPATION IN PLANNING

If a change only affects you, you can plan for it yourself. If, however, change affects more than just you, you need to have participation in that planning process. Further, the diversity of that participation needs to mirror the range of those impacted by the change. This information can literally be the difference between implementation success and failure. Again, the strength comes from differing FORs. Plan for change that recognizes all FORs and includes actions to meet different needs encounters fewer problems and unforeseen events, is easier to gain support for, and is easier to implement because it minimizes the negative feedback received.

IDENTIFYING TRAINING AND COMMUNICATION

Make sure that your change implementation plan includes two critical parts. First, include any training targeted at knowledge or skills development that is necessary for your change agents *and* the change targets. If you are going to be asking agents or targets to do anything differently, you need to assure them that they will have the knowledge and skills necessary to do it. Unfortunately, many change efforts have landed on the rocks because targets were not part of the plan.

The second critical elements to include are communication and communication pathways. Change is a critical time for communication, thus it needs to be part of your plan. The forms or pathways you will be using during implementation should also be described. Whether those pathways already need to be created to support the change implementation, they must be part of the plan if they are to receive the necessary resources.

PLANNING NONCONTROLLED RESOURCES

It would be nice if we controlled all of the resources necessary to successfully implement a change. Unfortunately, in most of the change cases, we don't. For example, you may have identified that the company newsletter will be the main source of communication about change. The company newsletter, however, is probably controlled by another person, department, and manager. *You*, therefore, have no control over that resource, although it is vital to your effort. Noncontrolled resources need to be identified early in your planning process and the issues and possible solutions discussed. An ideal solution to noncontrolled resources is to make those who do control them part of your change leadership team. By doing so, you develop ownership in the change result in key people. If, however, one of your necessary resources is having supervisors cover the change in regular workplace meetings, it is impractical to have all supervisors on your change leadership team. In such a common case, avenues and options to gain "influence" over these noncontrolled resources need to be placed early in your change plan. Because the resources are noncontrolled, gaining influence may be challenging and may call for very nontraditional measures.

ACCOUNTABILITY AND PLANNING

In reality, planning and accountability should be a matched set, existing only together. In practice, however, they are not. For any plan to be successful, accountability must be locked to every planned action. A specific person must be attached to each task. A deadline should be established for each action. Accountability ensures that each action item will get done or someone specific will need to say why. Without assigned accountability you will only get delays and problems blamed on someone else.

SCOREKEEPING AND PLANNING

Consider implementation of change to be like a ball game. The period of implementation begins with the start of the game and ends with the final play. The purpose of any ball game is to win. What element is critical to knowing if one side wins or loses? The score. What is a critical piece of information each team needs to plot strategy and keep motivated during the game? The score. Merely going thought the actions of the ball game is never enough. Without a score to refer to, the outcome is meaningless. It is the same in implementation of change. The purpose is to successfully implement your change. You cannot expect to know when the game is over, whether you won or lost, or if you need to change your plan during implementation if you don't have some sort of measurement—what I call scorekeeping. Scorekeeping is a critical element of any change effort and must be included in your change plan. How will you measure your progress? What will you measure to check that critical actions are done? What will you measure to assure your sponsors that the change will have the expected results? What will you measure to tell you that it's time to pop the cork?

GETTING THE PLAN APPROVED

There are two entities that *must* approve your plan—your boss and any sponsors you might have. Like it or not, they are an integral part of your change effort. If the effort fails, they will also suffer the loss. Because of this "ownership" in the intended result, they have the right to have input into your implementation and once it is finished, they need to approve it. This is not just a ritual. It is necessary to ensure continued support, provision of resources, and their comfort. It also helps keep you in what Chuck Coonradt in *The Game of Work* calls the GOMB zone. It stands for Get Off My Back and refers to the portion of the field of play where you are not overly managed by your boss or your sponsors. You have a much better chance of playing in the GOMB zone and allowing your plan to be implemented if you seek approval first.

COMMUNICATING CRITICAL INFORMATION

The planning and implementation stage is no place to forget about the critical nature of communicating. We talked about some of these necessary pieces of information earlier. You now have another piece of critical information—your plan. Tell them what they will see. Change is more frightening if we don't know from day to day what we will see and experience. Communicating the plan also lets others judge how far along the implementation path they have traveled.

It is also an excellent time to recommunicate the benefits of the change. The planning process may have taken some time during the development and approval stages. The benefits are so important that you cannot assume that everyone will remember them correctly or the same way.

It is also a good time to repeat the true impacts of change. Everyone who will be affected by the change or will see the impacts needs to know they are coming. Whether the impacts are changes in work loads, the need to learn new skills, or different reporting structures, everyone needs to be aware of all the impacts anticipated in the change effort.

LISTENING FOR FEEDBACK

Feedback becomes critical information to a change champion and a change leadership team in the process. Only through feedback can you judge if you are accomplishing what you want or are finding problems, both identified and unforeseen, that need to be dealt with quickly. It is critical that you become a good and attentive listener.

Most auto mechanics have a special trick for diagnosing a problem. They know that if an engine is making a funny noise, it's very difficult to determine exactly where it is coming from, and equally important to know the source if they are going to fix it. The mechanic places one end of an ordinary broom handle on various places around the running engine and presses the other end against the temporal bone next to the ear. Noises coming through the broom handle are amplified and the problem area can be quickly identified. This would be a great trick for us to use when we listen for feedback during change. With all the positive and negative feedback, it too often sounds like the blending of hundreds of moving engine parts. Just like that master mechanic, it is our job to listen attentively to all the sounds and to be able to identify problems early, using feedback.

The need to listen to feedback is another reason why diversity in your change leadership team is so important. Because team members come from their own area of expertise, have much more in-depth knowledge of different areas in the organization and have their own FOR, they can hear feedback that you may lose in all the normal "hum" of change. Properly alerted to "listen" and being aware of their important role in receiving feedback, they can serve as that broom handle does to the master mechanic. Because, most often, the feedback is there—hearing it is the trick.

REVISING THE PLAN

A plan for change is a dynamic tool that is interconnected to implementation. It evolves as feedback or other things occur in the change implementation. There must then be an available option for revising the plan. The urgency, however, is much different once implementation has begun. Before implementation, there is a lot more room for taking one's time with revisions. Once implementation starts, however, that time-buffer disappears and there is a need for rapid actions via change revisions to the plan.

Part of your change planning process needs to itemize how those rapid plan revisions will be accomplished. If your change leadership team is small, this can be a group action. If, however, the team is large, include additional planning for rapid revisions to the change plan while including important FORs and maintaining team comfort.

BEGINNING IMPLEMENTATION

Implementation is the point in the change process where most failed change efforts begin. Following our traditionally American way of "just do it," many change efforts begin with implementation and just skip the pre-change stages and planning. This is especially true if upper management orders the change; they expect change to be immediate. Successful change efforts are built on firm foundations. Those foundations are direct results of the pre-change stages and planning.

You cannot make a change that catches others unaware. The Baltimore Colts leaving Baltimore for Indianapolis should make this point very well. People cannot just wake up and find things in their lives different. Sudden, unanticipated change causes shock waves and ill feelings that take a long time to heal. If you are going to make a change that impacts others and benefits the organization, you will have to deal with the fears and concerns now or at sometime in the future. Poorly chosen, you may be dealing with them for a long time. The secret to dealing with them now and according to your well strategized plan is visibility of the change and change process. That starts with communication. This should be the first action in your change implementation process. The greater the communication blitz, the smaller the shock wave you will get. Use many routes of communication—most assuredly including verbal exchanges—to get your change message out to those who will be impacted by the change.

The second step in your implementation should be training. If something will be different, knowledge and skills in the targets must be increased. Be specific about what knowledge and/or skills are necessary for those impacted by the change to be successful in the "new world," and about how you plan on providing that training. Training must precede expectation.

Third, if your change will be driven by some formal process such as procedures, policies, or programs, they need to be developed, reviewed, approved, communicated, and to become part of your training actions. It is always best if some formal mechanism drives a change because it is too easy, especially during transition, for people to return to their old ways. Formal mechanisms set the new course and chisel it in the concrete of your accepted organizational guidance documents. Acceptance and conformance are thereby much easier to administer if you use the organization's mechanism for standardizing practice.

HANDS-ON VERSUS HANDS-OFF IMPLEMENTATION

There will probably be two different aspects of implementation. Those actions over which you have complete, direct control—the hands-on items. There will probably also be a number of actions, perhaps most, over which you may have influence, but have no direct control. These hands-off items may include communication pathways, workplace meetings, supervisor contact and reinforcement, worker involvement, and management changes or worker paradigms. If you built additional control elements within your change leadership team, however, there may be a considerable portion of your planned actions that requires some nurturing and faith. This is a normal part of change implementation. Necessary hands-off actions provide an ideal opportunity for plan and scorekeeping visibility. If completion of those actions is highly visible, it serves as a scoreboard at a game. It draws attention, motivates others to action, and documents successful implementation.

REGULAR MONITORING OF IMPLEMENTATION

Measurement or scorekeeping is essential for the successful implementation of change. Ideally, your change plan should be adapted into a timeline (Gantt Chart-type format) and displayed. This is an excellent way to measure progress in your change effort. As planned actions are accomplished, the Gantt Chart shows the progress. Monitoring implementation is one of the major jobs of the change champion.

Regular measurement accomplishes five things. First, it keeps the change effort going. If no measurement is performed, it is easy to allow the process to come to a stop. Second, it serves as a reminder to those who are accountable for making the actions happen. If your measurement is highly visible and seen by upper management, the measurement will save you the bother of reminding them. Third, it allows those involved in the process or impacted by the change to visually note where they are in the change process; it serves as a means of communication. Fourth, it keeps your sponsors aware of progress or identifies where the effort is struggling. This not only gives sponsors a sense of comfort but tells them where additional or increased resources may be needed. Fifth, it is extremely motivational for you, your change leadership team, your boss, knowing he or she backed the right plan, those who are (or aren't) getting their accountable actions accomplished.

Regular monitoring of implementation against the plan is necessary for two other important reasons. First, it is a feedback process for your plan review. As you meet roadblocks, your efforts are not as successful as planned, or when unforeseen difficulties slow or stop progress, your plan review and revision process can go into action. Plan corrections can be made. New resources and ideas can be mobilized. New people can be assigned accountability. A number of corrective actions can take place as a direct result of monitoring your implementation. The second reason is to identify areas where additional, better, or new avenues of communication are needed. Communication is the lead horse in the charge. Communication needs can be quickly identified from measurement.

COMMUNICATING WITH SPONSORS

The need to keep the communication level high with your sponsors must be emphasized. Sponsors offer a significant commitment based on faith and they require continual assurances that they made the right decision in backing your plan. Prior to and during the implementation of change, assurances can only be provided through regular communication. Once given, sponsorship is *not* a guarantee. Like a garden, it requires continual care. This can only be provided by communicating with your sponsors regularly.

Sponsors don't necessarily want to hear the day-to-day details. A sponsor's time is valuable, so don't waste it. When significant things get accomplished, tell them. When you need additional help, guidance, or resources, tell them. When things are going well, tell them. When things aren't, although it is very difficult, tell them. If your implementation plan is fairly demanding, this communication automatically becomes routine.

Communication with sponsors can take many forms. Memos and e-mail are acceptable, but direct verbal communication is preferred. It provides opportunities for questions and for nonverbal communication. Both of these can be significant problems for retaining sponsor comfort if only one-way communication avenues such as memos and e-mail are chosen regularly.

KEEPING IMPLEMENTATION MOVING

Planning and implementation are synergistically connected. Implementation builds on the planning process. Planning feeds on successes and failures during implementation and the continual stream of feedback coming from the process. Planning is continually reviewed, revised, and validated. It guides the implementation process. The traditional concept of "just do it" cannot succeed in change. One could also argue that with the ever increasing need and demand for change, planning and implementation fill a continuum in today's world. As we move from one change to the next, from one phase to the next, change becomes part of our everyday existence.

6

THE TRANSITION STAGE: NO LAND IN SIGHT

On an airliner crossing the Pacific Ocean, I became mesmerized by the expanse of water and the endless small puffy clouds below. It seemed that the scene went on forever, unchanging. I began to wonder how someone else would feel seeing this same view from a much slower form of transportation. For example, this endless expanse could be a significant concern if you were alone on a sailboat on the water's surface. When you travel on land, you never seem to get that feeling; you always know that you can stop somewhere. On the ocean, there is no place to stop.

I knew that in a matter of a few short hours, I would not only see land, I would be on it. What existed below in a seemingly endless stream was merely a curiosity from 38,000 feet and at over 500 miles per hour. The altitude was secondary to the speed; it provided clarity to the view below. At a lower altitude, the view would be blurred, but I would still be assured of finding land. The velocity of travel, however, is the significant factor.

VIEWING THE TRANSITION STAGE AS A CHASM

In comparison, rapid change can be accomplished either in a short distance or at a high velocity. Transition from start to finish would be relatively quick, assured, and painless. Change that involves complex

Transition

transition, however, is very slow. Just like the ocean, transition seems to go on forever, often leaving you adrift and controlled by outside forces.

In the transition stage, negative aspects tend to be at a maximum, including the level of effort required, amount of resistance encountered, and frequency of covert communication. Sponsor support, resources, and optimism are at low levels. The message is old. Covert communication is at a constant roar. The demand for cheerleading is ever-present and emotions are at their highest level. Transition is the part of the change process where demand, fatigue, and resistance are at the highest levels, and energy and commitment are at their potential lowest levels. Because of these factors, transition can be the most difficult stage of the change process. It is literally the point where there is no land in sight.

A mid-sized Midwestern manufacturing facility had decided that a change was necessary to ensure their market viability. They needed to make employee participation a driving force in all aspects of the business.

To enable this, they would have to simplify the organization so that fewer boundaries to participation existed and a smoother path for empowerment was provided. Planning and communication were extensive. The change process was initiated and teams were formed to begin the reorganization process. What would the new organizational structure look like in this empowered company and how would it function? Two years later, the teams were still meeting, and struggling to determine the structure of the new organization. A seemingly clear vision had become an amorphous image that was in wide disagreement. Efforts of the teams were fragmented and headed in different directions. The plan had become so convoluted and confusing that it had to be discarded. They were adrift with no land in sight. They couldn't go back nor forward.

This company's journey toward change was long and complex, and they had not planned carefully for transition. Because the currents and storms are so strong in transition, they had swept the change effort off course into dangerous seas. An even greater challenge than the original one involved re-planning the change amidst disarray and confusion.

THE CRITICAL NATURE OF TRANSITION

Vulnerability. Change champions cannot forget that change can only be accomplished through voluntary or forced vulnerability. In other words, you must remove the barriers and the defense mechanisms so that change can occur. If the protective walls stay in place, change cannot occur. Removing these protective walls, makes people, programs, and organizations vulnerable. Vulnerability is unacceptable for an extended period. Transition is that period in change where vulnerability is supposed to be replaced with assuredness. This assuredness can have negative or positive consequences. If the progress of change is delayed, stalls, or moves backward in transition, the need to move past the vulnerable state becomes more urgent. If the vulnerable state cannot be escaped, disarray, and chaos often result.

Emotion. Emotion is that nonlogical part in all of us that defies definition or predictability. The emotional response cycle to change was discussed in *Total Quality for Safety and Health Professionals*. Beginning with anger, the emotional response follows a predictable cycle if it continues to acceptance. However, it seldom moves in one smooth, nonreversing, noncycling manner to completion. It is a confused cycle that is both predictable in its phases and nonpredictable in its movement and progress.

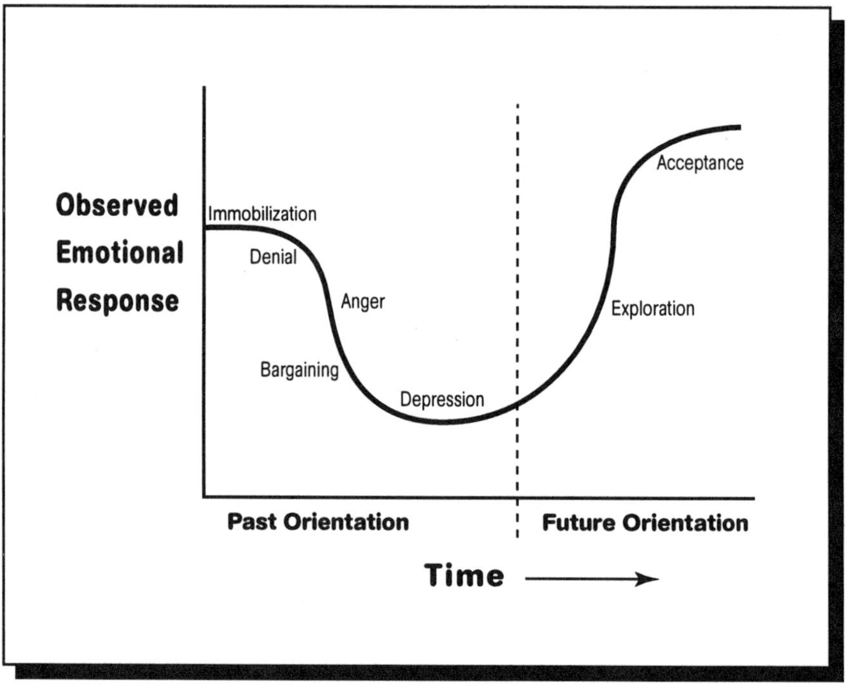

Figure 6.1. Emotional Response Cycle to Change.

As a review, the cycle of change includes the following stages: immobilization, denial, anger, bargaining, depression, exploration, and, finally, acceptance. The critical stages in transition fall between denial and exploration. The critical element is the orientation of those caught in the change effort. Are they looking back at what was prior to the change effort (usually longingly), or looking forward (usually without excitement), toward the possible future? Emotional response reaches it's lowest ebb in depression; it's the point at which one gives up. Only by moving through this point can one change from looking backward to exploring what could be in the future. It seems that at the change's lowest emotional point, one is nearest the turn toward victory.

A professional colleague wanted to move from being an employee to being a contractor with the same company. It was a voluntary outsourcing effort. He saw many advantages, both personally and financially. His employer also saw great advantages and cost savings. Both he and his employer were in agreement. Lawyers were brought in to draw up the contract which included noncompeting and conflict-of-interest clauses, no-cut verbiage, and rate guarantees that would protect both parties. The contract was finally written to both parties' satisfaction. My friend had excitedly inked his approval and the contract was sent to the General Manager for his signature. Three months later my friend was still waiting. The problem? An emotional response somewhere between anger and denial kept looping around in the General Manager's mind that simply wouldn't let him sign the document. It was very embarrassing for the company and frustrating for my friend.

The change process places everyone involved somewhere in that emotional response cycle. Change champions get a first-hand look at loops and nonprogress in emotional responses.

Giving Up. At no time in the change process is there a greater temptation to give up than during transition. Some mountain climbers, during a very long and difficult climb, never come out of their tents and simply freeze to death. Good runners understand that the transition stage of a race requires the most determination. It is the reason why some basketball games end in enormous score differences. Throughout history, heroic come-back efforts have occurred because effort in the transition stage is renewed. Transition is the time when giving up is most prevalent and at it's highest temptation.

Transition, therefore, is an opportunity for those with the uncommon ability to stay focused. These are individual qualities that are not as common as the temptation to just forget it. Since change is seldom accomplished by one individual, select those who can excel in the transition phase to be involved in your plan for change.

Perceptions. There is a saying in running that goes like this, "The most difficult part of a race is the middle." This speaks of perceptions during this critical part of transition. It is not a reach to think of transition as the middle of a long, hard race. You are so far from the start and have so far to the finish, conflicting perceptions are present. At this stage of the race, the pain is beginning to take its toll. Why would you continue to put yourself through this increasing agony? It's a problem of perception. In reality, in a race, you may only be a few short minutes from the start and to the finish, but an organizational or program change effort is much more long-lasting. Perception is instantaneous and locked in magnitude to the degree of disruption, not to time. In a race, pain and discomfort are reality. Time is only the period one expects to endure that agony.

Transition is a very difficult phase to endure. That's why perception is a major part of the critical nature of transition. Perception is rarely based on reality. Because perception is a mind-issue only, you have to rely on mental approaches to avoid problems. Faulty perception often causes others to forget the benefits of the change. This tendency is greatly dependent on the amount of disruption or pain the change inflicts. There are simple correlated target-needs that can reduce the level of disruption. We've talked about the importance of communication before. It is critical in transition. Making sure that all the "why" information is included in a constant roar of communication throughout the change effort is the best avoidance action to problems from perceptions.

KEEPING THE MESSAGE FRESH

Communication must be constant in a change effort, but the message must be kept fresh and interesting. We tend to look at communication as a qualitative, not a quantitative, action. This is an important difference. Qualitative only measures "yes or no," "on or off," whether information was conveyed (one way).

I had a friend who was employed as a Communication Specialist for a major American business. She was a professional at communications, but was unable to understand the difference between qualitative and quantitative communication. She mistakenly thought the

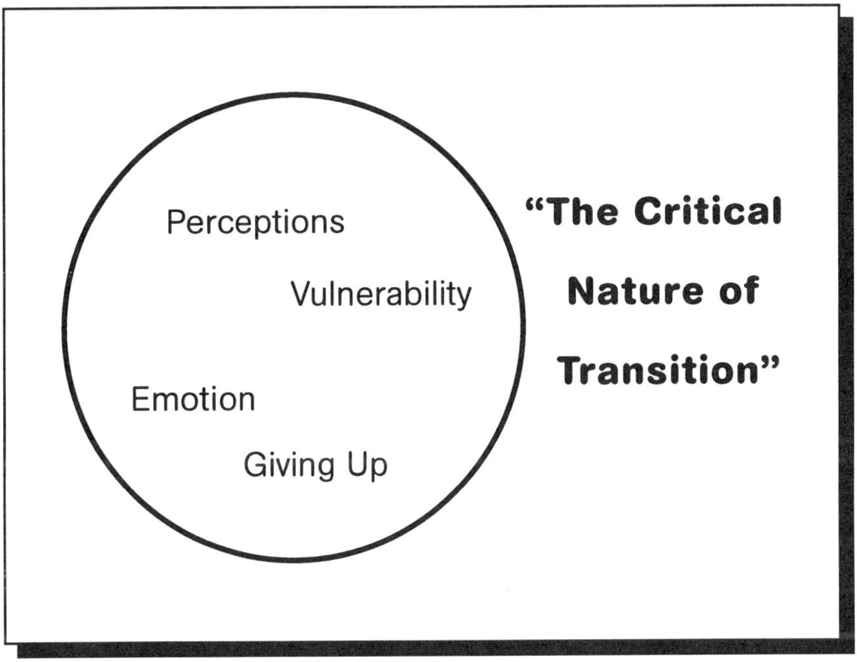

Figure 6.2. The Critical Nature of Transition.

difference involved clever writing or being able to tell something her way. Not so! I have another friend who has an undergraduate degree in com-munications, but works as a Safety Professional. He understands the difference all too well. It has little to do with how the information is pack-aged or conveyed; rather, it involves meeting the needs of the listeners.

An empathetic way of communicating necessary information looks through the "receiver's" eyes and ears. This "packaged" information answers four critical questions: 1) What information does the receiver want to know? 2) What is the best way to transmit this information? 3) Where is the best place to receive this information? 4) When is the information wanted? Effective communication establishes a connection between the transmitter and the receivers. This is the difference between qualitative and quantitative communication.

Note that a subtle difference in phrasing of the question represents a phase shift in communication from the transmitter to the receiver. We are programmed to view the world from *our* perspective. When we talk about traditional communication, we talk about communication *needs*. From a transmitter-oriented viewpoint, "What do they *need* to know?" For communication to be effective, the transmitter must look at communication from the receiver's orientation: "What do they *want* to know?" The first is an assumption based on a one-sided assessment that is too often power-based. The second is an approach toward communication.

Look Through the Receiver's View!

- What information is wanted?
- What way is best?
- Where is it best?
- When is it wanted?

This perspective is important if one is to keep communication fresh in a change effort. There are a few important truths about communication in change that clarify these points. First, *during the change, information needs, effective pathways, and best places, and times for communicating information also change.* The receiver's needs do not remain constant during the change process.

Second, *in a change process, communication resources and relationships change.* A change effort may increase or decrease the number of communication pathways. A newsletter may be discarded. An electronic means of communication may be added. A communication position may be added or removed. The reporting structure that you depended on for communication may be altered or removed. Change can also have two very different results on the relationship between the champion and sender (you), and the receivers. Your credibility will increase or decrease depending on the success (or failure) of the change process, the amount of disruption the change has brought, and the time the change effort has taken.

Third, *during the change process, the message may change.* If everything is going well, perhaps the message is a constant. If, however, modifications in the change plan or goal became necessary or unanticipated truths became known, the message will change accordingly.

Fourth, *if the message changes, tell the receivers.* A change in message, of course, is something that you know and they don't. That's the very reason why communicating changes is so important. But, too often we assume that just because the message has changed, even just a little, the receiver is somehow supposed to know that communication is needed. Not so. This comes from our orientation on communicating, from transmitter to receiver. The point here is if the message changes, get it out quickly and truthfully. In change, receivers *want* to know when the message changes.

Four pieces of information need to be communicated in a change effort: progress of the change, the road yet to be traveled, victories and failures to this point, and the benefits and importance of change. *Progress* is important to those caught up in change, contributing or cast in the current, or impacted by change. Others often assume that if nothing is communicated, no progress is being made. This false assumption can foster negative rumors and resistance. Any progress against the change implementation plan is important to all. There is a simple rule about communicating progress in change. Like the first step a toddler makes, every little step is important. Don't belittle *any* progress: even small steps forward should be communicated and celebrated.

The road yet to be traveled is another important piece of information to communicate regularly. When others concentrate on the distance from the start it may be destructive to your efforts to keep the eyes looking toward the new horizon. You should emphasize the distance that must still be traveled to the end of the road. Communicating this lets everyone prepare for what is to come.

Communicating the *victories and failures* thus far in the change effort is also important. No one expects only victories. We often expect more failures than successes. Thus, communicating those victories or milestones reached is more important than talking in detail about the failures. However, the failures should be acknowledged, honestly, so that everyone can learn from them. The concept here is learn from your stumbling so that they are not repeated. Victories, however, are the elements from which momentum and renewed spirit are derived.

Memory grows dim in change, especially if the change effort is long or has great impact. As part of your communication plan, include regular reminder messages. Why was the change a good idea for the organization or the program? What are the expected benefits and what will they look like? Why is this change important for the organization or program? These reminders of the *benefits and importance of the change* should be the starting messages for every regular communication. They are the anchors for your change. They are the fingerprints of the effort. They make it possible for others to be able to recognize victories or when the war is won.

KEEPING SCORE

Change is a busy time for all involved in it's birth, especially change champions and change agents. It is also a time when details are most often forgotten or de-emphasized. The scorekeeping action that is used to track the change progress and improvement objectives often falls victim to forgetfulness. This is a critical mistake. An example from the sports world will illustrate the importance of scorekeeping.

Suppose that the National League decided that more home runs were needed to increase the attendance at the games. They decided to change to a better-designed ball, one that would be easier to hit out of the park. They did their pre-change work, planned and then implemented the change, and had an active communication program to keep the message fresh. But, in all their effort to successfully get acceptance from the clubs, the players, and the fans, they stopped keeping track of the number of home runs hit and the attendance at ball games. What would be the result of such a scorekeeping blunder? First, acceptance would be more difficult to achieve because progress would be based solely on each person's perspective. This would result in increased resistance. Second, motivation would decrease, even at the National League office, because too much effort would be expended dealing with the resistance. Third, at the end (if the push for change outlasted the resistance), no one would know if the original goals of more attendance and home runs were accomplished. Without scorekeeping in transition, the process would require more effort and the end result would be unknown.

CORRECTING THE PATH AS NECESSARY

Even the best planning effort needs to be adjusted. This is no reflection on the quality of the initial planning process. We simply are not able to plan for everything we will encounter, every resistance or roadblock we will meet, every cultural icon we must combat, every person with organizational power whom we threaten by the change, etc. Conversely, sometimes we plan too well and are surprised when the path meets with virtually no resistance. Transition is the stage where we gain great wisdom from actual experience. We need to keep the implementation path dynamic.

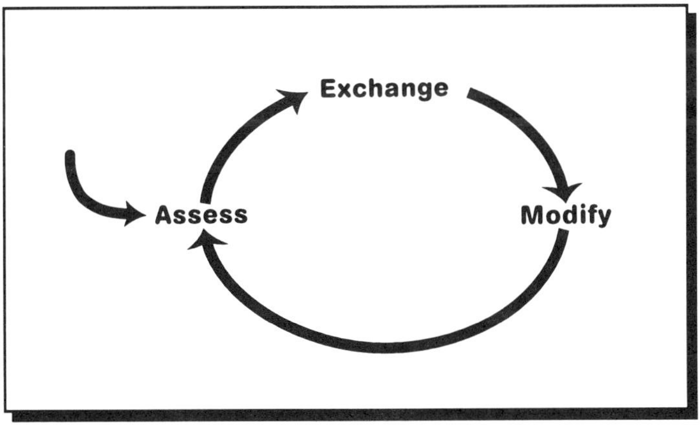

Figure 6.3. The Evaluation and Modification Cycle.

The cycle shown is the process for regularly evaluating and modifying a change implementation plan. There are three elements or steps in the cycle—assess, exchange, and modify. *Assess* is the step where you focus on what might need to be modified in the plan. Assessment should be a scheduled event in the change implementation plan. Scheduling an assessment of the plan prevents becoming so caught up in the implementation that critical inputs or problems are not seen. This step encourages you to compare the plan against the preplanning evaluations and current level of progress, and ask three questions. 1) What unexpected things have occurred? 2) In what areas are we ahead or behind schedule? 3) Why has this occurred? In assessment, unanticipated change issues are formally identified and a preliminary search for the cause(s) is attempted.

This regularly scheduled assessment step leads to the *exchange or feedback* step. In this step, you should ask three additional questions. 1) We think we know what the new and/or unanticipated change issue(s) is, are we correct? 2) What is the true and/or underlying cause of the issue(s)? 3) How can we deal positively with the issue(s)? These questions *must* be asked to those outside the change implementation team. The feedback from these questions is not only extremely valuable to the potential success of your change effort, but there is a power in asking for help, even if it is from the enemy. It builds bridges, encourages understanding of different FORs, and creates an opportunity for contribution to the change effort.

The last and obvious step in this cycle is to *modify* the change implementation plan. In practice, however, it is often the only step taken. It is a reversion to the insufficient planning practice we traditionally follow, which normally fails. The modifications made *must* be in response to the answers to the assessment and exchange step questions if they are to be beneficial to the implementation plan. The cycle continues beyond the modification by returning the assessment portion of the cycle.

Remember two critical times when you modify your plan. First, inform your sponsor (and your boss, if he/she isn't the change sponsor) about the change(s). Because they provide resources and through their power, authorize the change, they need to know what is changing and why. Second, modifying the plan will most likely or to whom information is communicated. Your plan modification should also include modifications to the communication effort.

KEEPING SPONSORS UP TO DATE

Transition is a critical stage for keeping your change sponsors informed. A sponsor is in a totally reactive role. When you need resources, authority, or support, you ask and they provide. Also, as resistance surfaces or the use of power is necessary, the sponsor reacts according to the commitment made. These are reactive aspects only. If communication concerning a change effort lags or ceases, all the sponsor hears is the sounds of resistance, which can be discouraging. Keeping a sponsor up to date includes constant communication about how the effort is going. Additionally, the sponsor must be made aware of and support any changes in the implementation plan.

DEALING WITH SPONSORS' PERCEPTIONS AND FEARS

When he was 16, my son came to me one day and asked, "Can I borrow your car to take my date and another couple to the school dance?" My pulse quickened. My mind raced as I tried to find a convenient way of saying "Gee, I don't think that's a good idea." This response from me, of course, would have been devastating to my son—it was purely emotional. I had complete faith in his driving. What could I say? Why was I feeling this high level of anxiety?

In fact, my son was asking for me to sponsor this change. Change always has a certain amount of risk assigned to it and by it's definition, it changes predictability. I was anxious, because by saying yes, I was stepping out on that limb of risk and had no direct control over the outcome. It was literally a step in faith. This is what anxiety is all about. Looking at anxiety in this light, however, it is understandable why a sponsor for your program change might also have anxieties.

Anxiety in your sponsors is a result of perceptions and fears. *Perceptions* come from two areas. First, what we don't know and feel, we have to guess. Our so-called intuitive sides are often absent . So we guess and call it intuitive. Sponsors are no different.

Perceptions also come from history. Having seen something once, we assume it has always been that way. This is called observation management. Having seen something twice or three times, we make the same assumption. This is called judgmental management. Our knowledge can also be based on fact. Usually, this is from long-term dealings or from direct measurement. This form of management, however, is becoming rarer because of the widening scope of responsibilities (via the flattening organizations), our ever-increasing transient employment practices and shifting organizational structures, and the need to constantly do more with less.

Fears have many roots and causes, but they're generally a reaction to the unknown. Just like your reaction as a young child to that monster that you were sure lived in your bedroom closet, our fears change as we grow, but our reactions to them continue. For this reason alone, change champions can not ignore the natural fears of their sponsors.

If constant and dependable support from our sponsors is critical to the success of your change effort, how then do you deal with these sponsors' perceptions and fears that can so easily derail our efforts? In a word, communication—regular, complete, honest, and repetitive communication.

In change, especially to a sponsor who has stuck his or her neck out in support of your change, a lull in communications can be deafening. No news is *always* perceived as bad news. Regularly scheduled progress meetings with sponsors can be extremely important for dealing with their perspectives and fears.

But, this constant nurturing of sponsor support must be planned and scheduled. In transition where we tend to battle everything all at once, we cannot forget to communicate with our sponsors. This is a critical lesson for change champions. If you forget this, your other battles in the trenches of change will seem like child's play and become secondary in importance when sponsor perceptions and fears cause the support rug to be yanked out from underneath you and your change effort.

USING ACTIVE LISTENING SKILLS

Communication is a two-way operation. Probably the single most important reason why we communicate poorly is that we spend too much time transmitting and not enough (if at all) receiving. Have you ever been "tuned out" by another person? How did it make you feel? From this perspective it is easy to see how listening is important, from *our* FOR. But, we fail to see the importance of our listening from *their* FOR.

In the change effort, active listening is a powerful tool. Change champions are supposed to talk a lot, but often forget to listen. Actively listening is more than hearing or reading, it hooks our sense receptors to true communication.

There is another potential flaw we have when it comes to listening. Depending on personality and training, each of us has a built-in listening meter, that functions like a clock. When the meter expires, we stop listening. For some of us, this period of listening can be measured in nanoseconds. Unfortunately, during the last portion of our meter's time period, our minds begin to move out of receiving mode and into the "let's formulate what I am going to say once this person stops talking" mode. It is natural for us to focus on our own perspective.

Active listening keeps the receiving senses connected to the brain and blocks our natural tendencies. It takes training and practice to become proficient at this skill.

But, in transition, it is critical that we actively listen well. Why? How can we know how the change effort is going, what peoples' (including sponsors') concerns or communication wants are, or keep communication active and progressive if we don't listen? We can't. During the transition stage is when unanticipated information is brought to the surface. If we don't hear it, we can't respond to it in our change planning process and meet the communication desire of others.

DEALING WITH RESISTANCE

Transition is the phase during which resistance can reach its highest level—a result of the time orientation of those affected by the change process. Referring to the emotional response to change illustration (introduced earlier in this chapter), this time orientation is shown on the x-axis. The trick to successfully implementing change is to drive the emotional response through the past orientation to the future. The past orientation portion of the response curve is filled with negative energies and responses. Why can't things be the way they were? With all the disruption caused by this change effort, what was really wrong with the old way? This orientation creates resistance.

The emotional response hits bottom in the depression stage. This is the point where those impacted by change literally give up and quit resisting, but the stage is necessary if their time orientation is to change to looking toward the future. Only through changing to a future orientation can one begin to explore the new prospects and, thereby, elevate the emotional response through discovery.

Knowing why resistance occurs and where it comes from, and understanding that maximum resistance occurs in transition, how can you effectively deal with it? You deal with it by avoiding most of it in the first place. The amount of resistance is dependent on the magnitude of the emotional range, and to a lesser extent, the length of time required by a change effort. The time a change effort takes is secondary to how far the emotional response falls from before the change to when the change targets bottom-out and give up. Effective pre-change planning and communication minimize resistance in the transition stage.

Once you know that even the smoothest change effort will certainly have resistance, how can effective change champion deal with it? First, *never minimize or ignore resistance.* Resistance that is ignored, like a snow ball rolling down a snow-covered mountain slope, only picks up speed and gets bigger. Second, *confront resistance positively and with empathy.* If you deal with resistance as something dark and diseased, you will add fuel to the resistance fire and encourage it to spread. Third, *don't back down.* How you deal with resistance is a mark of your character and a test of your change vision. If you back down a little, those resisting will continue to "bargain," thinking that your change vision is negotiable. By backing down, you feed the natural cycle inside the past-orientation of bargaining to denial to anger. If you get stuck in this past orientation cycle, your change effort may never advance. Fourth, *communicate, communicate, communicate.* Communication as a means of dealing with resistance is critical. But at no time is active listening more important to communication than when dealing with resistance in transition.

> **Dealing Effectively with Resistance in Transition**
>
> - Never minimize or ignore resistance
> - Confront resistance positively and with empathy
> - Don't back down
> - Communicate, communicate, communicate

"FEEDING" THE CHANGE AGENTS

Those who are in the trenches implementing the change—the change agents—are indispensable to any change effort. Without them, change cannot occur. Too often, though, their critical role in change is overlooked or forgotten—that is, until the wheels come off the change effort. Then, they are rocketed to the head of the blame list.

Since change leaders are critical to change success, they deserve and need the skills and tools necessary to implement change (people skills, authority, communication skills) and they need to be communicated with regularly. In transition, however, their need for skills, tools, authority, and communication is at its highest.

"Feeding" the change agents refers to meeting these four critical needs—skills, tools, authority, and communication. Skills and tools for implementing change should be part of your plan. Furthermore, as conditions change or new tools or skills are identified, those need to be added to your plan. Authority, if you've done your job keeping your sponsors informed and comfortable, should also be taken care of. The communication aspect, however, is your job as change champion. They must never be left unaware of what is going on. If they are, they will quickly become ineffective as change agents or lose interest in the change effort altogether.

KEEPING COMMUNICATION OVERT

There's a saying in management, "If it isn't visible, you can't manage it." This is one of the driving forces in management-by-measurement, in which performance and results are routinely measured and displayed. Visible, they can be managed. This saying also communication in the change process. Paraphrased, it means, "Once communication hits the grapevine, you can't manage it."

The grapevine is the age-old rumor path that isn't out in the open. It's generally a lot quicker than the "official word," and much more creative and emotionally charged. Occasionally, communication surfaces, usually at the most inopportune and potentially destructive times. But its normal existence is as covert communication.

Keeping communication overt, visible, and manageable is an important lesson for change efforts. It is critical in transition. Covert communication is rooted in emotional response. Normally this response is not positive and comes in the past-orientation portion of the emotional response curve. It exists because overt information is not available, not timely, or not believed.

Change champions must understand their roles in the communication process.

1) In change, communication must be available to meet the change targets' desires for information. As we introduced earlier, this has little to do with what the transmitter thinks is needed.

2) Communication in change must be rapid. If you wait for information to be officially approved, it's too late.

3) There must be faith in what you say, which is based on the credibility of what you or the organization has communicated.

4) Communication must be "friendly" and convenient if it is to be absorbed. Memos are generally a very poor choice for communication about a change effort. E-mail allows too many assumptions to be made. Spoken communication, direct or trickled down through the change agents or other rapid, mass means of communication, such as electronic message boards or work place postings, can be very effective.

KEEPING THE FAITH

I've always loved the classic motion pictures, *The Ten Commandments* and *Ben Hur*, with Charlton Heston as Moses and Ben Hur, respectively. The historical accuracy or significance of the plots are not relevant. These are stories about keeping the faith through adversity, resistance, and bad luck. A strong hero fights adversity and ultimately succeeds. The hero is larger than the opposition or roadblocks thrown in his path.

Even though none of us bear any resemblance to these characters, an easy parallel can be made in reference to our roles as change champions in the transition stage of the effort. None of these hero figures ever gave up. A recent beer ad on television illustrates this point very well. An older Charlton Heston is standing in a small group of people trading stories. He says, "And he said" (meaning the picture's director), "Just stay in the chariot, Chuck. I guarantee you're going to win." The ad ended with a close-up, "True story!" says Heston. This is a simple but direct lesson for all change champions. Our challenge, especially in transition where resistance is at its highest and energy and resolve its lowest, is to stay in the chariot. This doesn't mean that we have to be super human, exceptionally creative, master planners, hyper-energetic implementors or even great leaders. As change champions, we simply must keep the faith and not give up the fight.

TRANSITION CHALLENGES IN A CHANGE TO TOTAL QUALITY

Because total quality concepts are much different from our traditional management methods and ways of thinking, some challenges exist in transition that we need to be aware of and prepare for. Four simple words illustrate these challenges—*slow, flow, mo,* and *show*.

The speed of change is *slow* in total quality. This is counter to our traditional American expectation of and demand for light-speed change. The two words we use are innovation and automation. Everyone wants to be known as being the innovator of the next new idea. Automation, oddly enough, is both sought after and feared. It is sought as a rapid method of reducing costs and improving production capacity. It is feared from the management side because it comes with a high price tag, and from the labor side, because it usually eliminates jobs. Automation is an acceptable method of improvement in American business because it matches our expectation of rapid change.

We Americans also have a trait (or lack of a skill) that is not only traditional, it is expected of us. We are impatient. Not just a little, we are highly impatient. If a plane we are flying on is late leaving, listen to us grumble. We cut-in or come back another time if we have to stand in a line. When our children do something wrong, we expect correct behavior *now*. A problem surfaces in the production line, it has to be fixed immediately. We look at slow change as foot dragging or a bad attitude. We expect changes *NOW!*

The total quality concept emphasizes that a slow change ultimately lasts longer. This is contrary to what we traditionally expect and the way Americans are programmed. This conflict presents a very large challenge to champions of change to total quality. But the problem usually doesn't surface until the transition phase. This problem occurs even if you have made an extra effort to stress that there will be slow transformation. Open avenues for communication from target to agents and champion(s) are critical to effectively dealing with this issue and continuing the message that total quality is different than what we traditionally expect.

Flow refers to a change in energy orientation. Traditionally, energy is considered the responsibility of the top management of an organization or program. Energy is a leadership issue. In a change to total quality, the leader's role is critical to success. But, in a total quality program, effective team energy is percolated up through the organization from the bottom. Energy is fueled by participation in the process. The problems that surface from flow in transition are brought about when the leader begins to retract energy supplements to the team and expects the energy orientation to change. The reaction is not unlike weaning a youngster. No mother looks forward to the weaning process because it generates a great deal of resistance disagreeable behavior.

But this change in orientation is important to a total quality program. It is the source of power that drives a total quality program far beyond what traditional top-down organizations and programs can yield.

Planning for and dealing effectively with this transition point requires communication and patience.

Mo refers to momentum. Keeping a change effort going through a long or slow transition period is always a challenge. In a conversion to total quality, this stage can be especially challenging. Why? Because it is slow in progression and advancement, and the energy orientation is from the bottom up. Because of these significant differences, momentum in change needs to be fed by specialized means—measurement. In fact, change to total quality is greatly dependent on the quality of the change measurement.

For this reason, total quality change champions must not underestimate the need for measurement. Measurement provides the assuredness that continual progress is being made during the slow process. Since energy is dependent on the participation of the masses, mass and individual assessment of the results are required for continued participation or momentum. These two assessments can only be provided by measurement. Measurement is critical to successfully implementing total quality and maintaining "mo" in transition.

In each of these three issues—slow, flow and mo—there is a common thread that is so important, it is set aside for emphasis in *show*. This common thread is communication. Communication was critical in effectively dealing with the natural resistance to the slow progress of a change to total quality. Communication was fundamental in helping spawn the change in energy orientation from leader to participants. And communication was a critical element in the through visible measurement of the total quality change effort. Obviously, it is impossible to show something that is not yet come about. This "show me" requirement can only be provided by detailed communication.

SUMMARY

Transition is the crossroads of the change process—when it moves from what was to what will be. The difficulties involved in transition are dependent on the velocity of that change effort and the emotional response (delta) of those impacted by change. Because of this, transition is seldom smooth. It is the stage when resistance reaches its maximum, memories of the change vision become the most cloudy, frustration is at its peak, energy is at a low ebb, resources are spent or in adequate, and commitment wanes.

The importance of the change champion is at an apex in transition. The champion role includes being an enthusiastic cheerleader, being prepared for roadblocks and resistance, planning and modifying the plan as necessary, measuring the change effort, and, of course, communicating. Think of a change champion as the sailor who must steer the ship through the troubled waters of transition. If the voyage to the future is to be successful, the champion must perform the critical duties under all conditions.

7

The Celebration Stage: We Made It!

Years back, I was an avid pro-football fan. I knew the players, the teams, the strengths and weaknesses of each, and the schedules. Today, however, I seem to find other, more important things to do instead of watching football.

One of the reasons I stopped watching was the institution of the "excessive celebration" rule which says it is okay to celebrate a little when you score a touchdown, but not too much. Granted, over the years I've seen some pretty wild end zone dances and acrobatics. But, first, what is wrong with whooping it up after you've scored on the best defensive players in the league? Second, at what level does celebration become excessive? Personally, if I had just scored on Deion Sanders, I would feel that I deserved a major celebration, perhaps I'd even invite a few of my friends to join me.

Some say that I celebrate too quickly and too well. Conversely, most others require a monumental accomplishment in order to feel that a celebration is warranted. I disagree!

But this high accomplishment requirement and low celebration level are traditional. I suppose that opening champagne to celebrate finding a quarter under the cushions of your sofa would be considered to be socially unacceptable. Most of us are missing a lot of fun.

Celebration

CELEBRATION AND CHANGE

> Rule 1: There must be a nonnegotiable commitment to celebrating.

Celebration and change implementation are important to each other. Celebration is nothing more than an extraordinary activity that acknowledges that some significant event has occurred, activity is completed, or effort is successfully ended. It is a high-spirited closing activity. Celebrations have two purposes. They announce the end of something or that an important milestone has been reached and they motivate us for the next effort, challenge, or change. Celebrations are both past- and future-oriented.

An organization went through a long, difficult change effort. Because they were more proactive than most companies, they had chosen to change when they saw a real economic advantage in doing so. This was far from the pain trigger level for change. The change effort took a lot of participation, coordination, planning, and communication. But, finally, it was completed. The measurement that management used to indicate when the change was successfully implemented showed all the anticipated results. But, like most companies caught in today's tumultuous change environment, the organization was already focusing on the next change. Moving directly to the next change effort, they forgot to celebrate the one just completed. A short communication was sent to "All Employees" announcing that the change was successful and included a brief "Thank You" and "Keep up the good work." One month later, a strange but predictable thing happened. The subsequent change effort was abruptly aborted due to lack of participation. A dismal and quick failure followed what was known to management as a splendid success.

Why did this happen? Why was it so predictable? Because the organization's management forgot the importance of celebrating, three consequences occurred. Because of the lack of closure, the first consequence was that those who had participated in the first change effort were left unsure if the change was successful or completed. After all, if it wasn't worth celebrating, did they succeed, was their effort in vain, or was the change necessary in the first place? Second, there was no breather, a time to regroup before the next effort began. This lack of a break made the efforts run together into what seemed like a very long transition stage. Third, a lack of celebration and acknowledgment of effort and success left the motivation tank empty for the next change effort. So, rather than beginning charged up, the next change began with reserve energy. As a result of these consequences, the next change was predictably doomed before the vision ever got off the ground.

Celebration doesn't have to be reserved for the end of the game, either. When do football teams and individual players celebrate? They do so whenever any significant event occurs. For example, you can observe celebration activities whenever a quarterback is sacked, a running back is thrown for a loss, a pass reception is broken up, a touchdown or field goal is scored, a long pass is completed, a hard hit is made, a fumble is recovered, a fourth down is converted, a long running play is made, a pass is blocked at the line of scrimmage, a punt is downed before going into the end zone, a deceptive play succeeds in fooling the opposing team, just to name some of the more familiar reasons or successes. The point is, you don't have to wait until the end of the game to celebrate.

ACCOMPLISHMENT VERSUS RATIONALIZATION

> Rule 2: Celebrate accomplishments only.

Now that the importance of celebration is understood, we should distinguish between accomplishment and rationalization. Chuck Coonradt in his book *The Game of Work* videotape training program, puts it this way. "When you bring your performance up to meet your goal, that's accomplishment. However, when you bring your goal down to meet your performance, that's rationalization."

A well-known college basketball coach was hired by a prestigious university after their fine basketball program had fallen onto tough times. They had hired and fired other coaches who had not been able to turn the program around. The team's record was having significant impact on their ability to attract and sign star players and to attract fans. Unfortunately, they had heavy competition in their conference and in their state. The new coach, however, felt that a turnaround was truly achievable. At the press conference announcing his hiring, he openly boasted of how he would turn the team around in just five years. He definitely had high goals for the team and it excited everyone at the school including those who would work hard to see it happen, and the boosters who would support the changes.

His first year was a disappointment like the previous ones, but the lofty goals were still openly conveyed. The turnaround program was reported to be "on schedule." The second year was better and the third was better yet. The win:loss record changed from 6:20 in year one to 16:14 in year three. During the fourth year, the team started off with six straight wins. But those who kept a very active account of progress, including some of the school's wealthy boosters, began to notice some disturbing things. First, no star athletes had been signed under the new coach's system; they were still being signed by competing schools. Second, the number of teams the school had played with relatively poor records had increased from three in the first year to 12 in the fourth year. Some of the teams with which the university had bitter but long-standing rivalries were being played less often and some were not even scheduled in the fifth year. The team was winning more and would undoubtedly do even better in the fifth year. But the reason for this success wasn't due to a higher level of play by the university's basketball team. It was due to a lower level of competition. Consequently, the coach's contract was ended a year early and a search began for his replacement.

This is an example of rationalization. Instead of setting a high mark on the wall and improving the performance to reach the goal, the performance level was maintained and the mark was lowered. This story has direct application to change efforts. It is common in change that when the going gets tough, the goal is lowered or the vision is modified. Celebrating rationalization rather than the achievement causes confusion within the organization.

There is another face of rationalization: when you focus on the journey and lose sight of the goal. Let's say that your goal is to travel from New York to Miami. You prepare for your trip with careful planning. However, during your trip south, Miami is suddenly threatened by a hurricane. Your trip is repeatedly scuttled by threatening skies, blustery and strong winds, rain, and heavy traffic from those leaving the storm target. You turn back or take refuge along the path. It is frustrating. Finally, someone gives you an alternative. "Why don't you go to New Orleans. There's no storm there." So, you go to New Orleans. The question is, do you celebrate making your goal and completing the tough journey?

Applying this example to change efforts, it is too convenient in tough transition stages to accept less than your final goal for change. This is rationalization not accomplishment. Celebrating rationalization tends to poison other change efforts that might be taken in the future and may be fatal to any effort for rallying the troupes again.

HOW DO YOU KNOW YOU'VE MADE IT?

> Rule 3: Quality measurement tells us when to celebrate.

No one begins a trip anticipating that he or she will only get 65% of the way there or end up somewhere else. But, in the case of a physical change, such as traveling to another site or city, it is easy to know if you've arrived. In process-oriented or management-system-oriented changes like total quality, it is much more difficult to know when you've crossed the finish line.

This is where *measurement* becomes critical. The quality of the measurement is critical. All measurement is *not* created equal. Careful planning of what measurement will be used, how input will be acquired, and how the scorekeeping will be made visible and kept up to date, are important factors. It is acceptable if you modify your plan as you learn or react to unanticipated events, resistance, and roadblocks. It is not acceptable for you to change measurement criteria in midstream; confusion is the result.

Whenever change is measured, an improvement or attainment level must be established. High quality measurement systems will track the effort toward this level until victory can be proclaimed. This is the time to celebrate. But there needs to be some allowance in your measurement system for "performance noise," the normal fluctuation of performance levels. This normally happens toward the end of the journey. It's important to include the agreed upon performance noise expectation in your communication so that momentum doesn't drop so near success. Those impacted by the change effort may not understand the "noise" concept.

For change efforts where a long or difficult journey is anticipated, it is also important to include milestones in the measurement system as celebration points along the path. Measurement can be your best score card, but if the goal remains too distant, the ever-present measurement system can also be a demotivator.

The use of consistent and high quality measurements can avoid the pitfall of rationalization. The target is always kept in sight during measurement and it cannot be altered easily.

WHY IS CELEBRATING DIFFICULT FOR US?

> Rule 4: Celebrations require our "ties" to be loosened.

Most of us tend to be very serious about life. This is especially true in business. From our childhood, we are taught to be serious. Adults advise children to: "Quit kidding around. Start acting like an adult. Being an adult is a serious thing. Vacations are a society-approved mechanism to let our hair down and have some fun. Certain weekend recreational activities are other examples of society-approved fun. But notice something important about these "approved" methods—none of them are present.

If we were to isolate one part of our lives, one place that is serious, it would definitely be at work. Why is this important to our discussion of celebrating? You simply cannot celebrate and remain serious. You've got to loosen your tie, take your shoes off, let your hair down, put a smile on, to celebrate.

After successfully guiding a company through a difficult change effort, I was invited to attend their all-employee celebration. I took a can of whipping cream. When I arrived, it was a very dignified (and dull) activity. The Plant Manager was there in his tie. He was over in a corner having serious discussions with other members of his staff. The employees were sitting quietly across the room. Cupping my right hand,

I filled it about two-thirds full of whipping cream. As I put the whipping cream can in my back pocket, I walked up to the Plant Manager, presented my right hand (palm down) in a gesture to shake hands, and said, "Who died?" Confused, the Plant Manager went to take my hand in his. I caught the back of his hand with left hand and sandwiched it in a firm, mushy, dripping, friendly handshake. I had placed him at a sudden decision point. He could choose to make a scene in the positive or negative sense. His shocked eyes met my smiling face and he broke out in laughter. The true celebration began at that moment.

I knew that creating an environment for celebration would be difficult for him, so I took the whipping cream. There was a lot of investment in this organization's culture—management versus labor. Some significant act or activity that breaks down nonphysical, society-approved, organizationally invested barriers was needed. People who conduct training sessions recognize a significant tie between having fun and learning. A significant tie also exists between having fun and celebrating. It's an emotional, barrier-free relationship.

It's too bad that each of us doesn't have a switch on the back of our necks that allow us to turn off the business face and turn on the celebration face. Some of us are so invested in our business face, that celebrating is detached and formal, at best. We must recognize that celebrating is not easy for the great majority of us, especially those higher in organizations who have invested heavily in being serious. Celebrations must also be derived from an emotional involvement for them to be rewarding and motivating.

> Rule 5: Comfort about celebrating may require that we reassess some of our paradigms.

Our reluctance to celebrate is also related to our business history. Those who are considered part of the "old guard" came up through the ranks in a business environment that was static. The manufacturing processes and machines changed little and the management philosophy and practices changed less. Predictability was the key job element. In this business environment, celebrating was out of the question. A job was a job, not something that needed to be rewarding or entertaining.

In reality, today's business environment is in constant change. Logically, it makes little sense to connect our historical ban on celebrating at work to this new changing environment. In fact, though, we do. It is a pattern that goes unquestioned.

A friend of mine watched a movie which starred two well-known "macho" male actors dressing and acting like women. As he watched, he felt very uneasy and couldn't pay more than superficial attention. "Wait a minute," he said to himself. "Why am I having this reaction to this movie? The acting is good. The story line is very entertaining. He became aware that it was all about his paradigms. His paradigm concerned these actors in their stereotypical roles as "macho" characters. Their playing roles at the opposite end of the spectrum challenged his paradigm. Once he recognized this, he stretched the boundary of his paradigm and began to enjoy the movie.

This is a clear example of how our historically derived paradigms can unconsciously affect our behavior. Once we recognize that paradigms are only self-instituted boundaries, we can easily and effectively change them.

DON'T CELEBRATE TOO SOON

> Rule 6: A celebration is not necessarily a party and a party is not necessarily a celebration.

If we understand that the celebration is really the official final step of a change effort, it becomes something that we really want to get to. No one wants to be caught in the transition stage forever. However, there can be dangers and warnings associated with celebrating. But there are a few that should be addressed. I call it the "too hard or too soon" trap.

I entered a work department one day and the office and shop areas were empty. But there was a lot of noise coming from in back of the building. I walked through the maze of equipment and every few seconds, a burst of laughter would echo my way as I approached the back door. I didn't know what was going on but it sounded good. Almost immediately upon going through the door I heard, "It's Dave," and a whole throng of merry-makers started toward me. Because I didn't know what was going on, I had a curious inner reaction—both excited and a little fearful.

"What the heck is going on?" I asked the smiling crowd. "We're celebrating," they responded. "Obviously," I said. "But what are you celebrating?" The face of the person closest to me went blank. "Hey Sam, what are we celebrating?"

Some people don't need an excuse to party. But a party should never masquerade as a celebration of an accomplishment.

> Rule 7: The magnitude of the celebration must match the magnitude of the accomplishment.

I heard of an office that celebrated an office rearrangement. Making significant changes to an office layout can be very disruptive and requires a great deal of participation. They enjoyed the celebration so much that all decided to bring food in on Friday afternoons. Pretty soon, they were having a party every Friday afternoon. A celebration must be attached to an accomplishment and it needs to compare in magnitude to that accomplishment.

> Rule 8: It (the change effort) ain't over 'til it's over.

An organizational change effort seemed to be locked in the transition stage. The measurement of completion showed that they were close to reaching their goal but at this point, even the smallest gains were a struggle. Everyone was frustrated, including the sponsors and change agents. The change implementation team struggled to make the necessary modifications to plans that finally would take them over the hump. Pressure to complete the effort was increasing from the sponsors. Was there an easy answer that would take them to their goal? One of the members of the change implementation team had an idea. "Let's have the celebration early. That will get everyone relaxed, improve all of our attitudes, and refuel the effort so that we can quickly make it the short rest of the way." Everyone wanted a solution so badly and no new ideas were on the table—this idea sounded plausible. The celebration had been planned for when the effort was moved up. The "We're So Close" celebration was fun and well attended.

Two of the three intended purposes of the celebration were obviously successful: The pressure to complete the change subsided and everyone was much more relaxed. The third purpose, refueling the effort, however, took a negative turn. Everyone stopped trying. The ground gained and displayed in the measurement quickly returned to pre-change numbers. The change was dead. So was the career of the change champion.

A celebration can provide the "carrot" at the end of the stick that drives change. It can be effectively used by change champions to motivate others to finish the effort. But it can also cause a great deal of pressure to be applied on the champion by sponsors, change agents, and targets alike. Especially in a long transition stage, there is tremendous pressure to "get it finished." This is where measurement and honesty in evaluating the status of the effort can help.

TYPES AND IMPACTS OF CELEBRATING

> Rule 9: Celebrating is a qualitative concept.

There is a natural misconception about celebrating. We think that a proper celebration requires a lot of expense and a respectable amount of time. Unfortunately, this is the misconception that management often has when you mention the word "celebration." It conjures up negative mental images of dollar signs, lost productivity, and borderline employee behavior.

Celebrations come in all shapes and sizes. They range from simple "job well done" handshakes in the lunchroom to extravagant parties. The natural reaction to the word "celebration" is not based on fact. But you have to be aware of misconceptions and misguided expectations when you start planning celebrations.

As we introduced in celebration rule #7, celebrations must not only match the magnitude of the accomplishment, they must also match the culture of the organization or group and any budgetary restrictions. But before we discount celebrating because upper management doesn't like it or we have no budget for it, there's a simple fact about celebrating that we need to know.

> Rule 10: Belief in the change measurement system must be built early to validate celebration.

The fact that you choose to acknowledge accomplishment *with* a celebration is more important than *how* much of a celebration is held. If the change is sponsored by a member of upper management, it also matters a great deal if that person cares enough to attend. These are the two major areas of celebration impact—if you acknowledge accomplishment (with a celebration) and if the major players (sponsors and champions) take part in the celebration. Celebration is a qualitative concept that must be supported by the major players taking part.

CELEBRATION AND TOTAL QUALITY

Tying the act of celebrating to a technical or physical change is easy. Doing the same to a philosophical, nontangible change can be difficult. We trust our senses: "The wall color used to be white; now it is light gray," "The machine noise in the shop used to make my ears hurt; now it doesn't." "The surface texture on the product used to be rough and the edges were sharp. With the new process, it is smooth all over." Change has indeed occurred and if the vision was properly sold and communicated, the result is successful.

Nontangible changes are different. If our trusted senses can't be used, we are left to our perceptions, beliefs, and fears. "I know that we are being asked for our help and thoughts a lot more than last year. And we seem to be working as a team rather than by management directive. We're also planning and measuring a lot more. But how do I know that this is not just window dressing? How do I know that my ideas and participation are really being used and aren't just another management sham?"

This is not only a lesson about linking celebration to change success, it is one in the believability of measurement when dealing with nontangible, philosophical change. Measurement must truly document successful implementation of nontangible change to be believable; particularly for changes to total quality. It follows that if success in nontangible change is difficult to measure, there exists some difficulty tying change completion to celebration.

Total quality is a dependent concept. It relies on a variety of concepts and components to work together. By themselves, participation, planning, and measurement are aspects of total quality, but they are not the entire concept. So it is with the process of change. You can't just pick up one of the stages and call it a successful strategy for implementing change. Celebration for example, cannot stand alone as *the* method for ensuring success in change. It is an integral part of the change process. But, so are planning, solicitation, and preparation. These are the areas from which the measurement of your change effort originate. In these stages are the keys for tying successful completion to believable measurement and to celebration. It is an entire process approach rather than a single concept or stage focus.

Does this mean that because tying everything to celebrating is difficult, celebrating is less important or it is less important that it be done well? No—quite the contrary. Because the change effort is nontangible or philosophical, it makes the tie between celebrating and the other stages even more important. A change effort has to eventually come to an end, successful or not. In the celebration stage, the group or organization agrees that the effort was successful and that everyone deserves to be praised. If success in nontangible aspects of change is not celebrated, it is assumed that the effort was a failure and that it wasn't that important in the first place. These assumptions are damaging. If we are unable to use our trusted senses to judge successful change implementation, measurement and celebration become critical. But the proper measurement parameters and data collection methods must be set early in the process. Everyone involved in the change process must buy-in early in the change process.

NOW THAT YOU'VE MADE IT TO TOTAL QUALITY

Celebrating is merely the act of acknowledging that you were successful in your change effort. A change to total quality has three significant rewards. First, roles and responsibilities are where they belong. Total quality reinforces the idea that safety and health responsibility must lie in the line structure and that each member must be held accountable for his or her responsibility. This must be documentable throughout the line structure. If first line management is held accountable for safety and health in their area, so must upper management. Management is responsible for the day-to-day observation and correction of safety and health problems or deficiencies. They are responsible for ensuring proper safety training and good communication (two-way), and for attaining high safety participation. The total quality system supports that the only way safety and health can be effective is where line authority is united with assigned holistic responsibility and accountability.

The safety and health group or specialized function provides a service to the line (customers) in total quality. Service is provided at the level requested, when and where it is needed. The safety and health function also serves to coordinate and standardize the efforts, provide participation tools and measurement, document policies and programs, and identify special needs and program emphasis. This is a customer oversight role, *not* an auditing-type function. The safety and health function is a friend of the responsible line structure because it helps them be successful and adds value to the entire organization.

Second, The organizational safety and health effort is powered by a high level or participation. Through true ownership in safety and health, the entire line structure, including all workers, provides day-to-day safety and health vigilance. Through unbridled empowerment, workers are free to take steps that are necessary or corrections that are warranted to ensure that a workplace is as safe and healthy as is practical. Participation is highly dependent on building safety and health knowledge and skills in all levels of the organization. Participation must also be measured to ensure full cooperation and continual improvement.

Third, total quality moves safety and health away from the traditional reactive mode into a proactive one. Operating from measurement, planning becomes the backbone of true accident and illness prevention. As new lessons are learned, they are analyzed for true causes, and then plugged into the planning process for ongoing prevention. As process or production improvements are conceptualized, proactive safety and health provides hazard evaluation and prevention strategies well ahead of implementation or operation, where workers are at risk. The safety and health program becomes process-oriented and less results-oriented because perfecting the process *always* improves the results.

THE SEED PRINCIPLE

You may have successfully implemented total quality in your program, *alone*. This can be a lonely, but productive adventure. In this case, there is always the "Seed Principle." The Seed Principle says that a good seed planted in any organization that produces good fruit, can eventually take over the rest of the field. Just as a corrupt influence can spoil the entire organization (the Weed Principle), a good influence can have the opposite effect. The Seed Principle usually takes longer, however.

If what you are doing works well and is well thought of in upper management, others naturally emulate your efforts. This is the Seed Principle in action.

SUMMARY

The celebration stage of the change process is more than an excuse to have a party. It must also be a part of the entire process. It is important because it signals a successful end to the change effort. The act of celebrating is the clutch that allows your organization to shift gears. But there are some simple rules concerning celebration. If followed, these rules should maximize the impact of celebrating success of your past effort, and give a considerable advantage on the next effort.

Rules for Celebrating Successful Change Implementation

1. There must be a nonnegotiable commitment to celebrating
2. Celebrate accomplishments only
3. Quality measurement tells us when to celebrate
4. Celebrations require our "ties" to be loosened
5. Comfort about celebrating may require that we reassess some of our paradigms
6. A celebration is not necessarily a party and a party is not necessarily a celebration
7. The magnitude of the celebration must match the magnitude of the accomplishment
8. It (the change effort) ain't over 'til it's over
9. Celebration is a qualitative concept
10. Belief in the change measurement system must be built early to validate celebration

8

The Continuous Improvement Stage: The Road to Excellence

Continuous improvement in American organizations is often misunderstood and misapplied. We try to Americanize the term when, in fact, it is very different from how we traditionally think. We are very comfortable with the quantum leap change concept. We call it "innovation." We use many *big* mechanisms (and expensive ones, too)

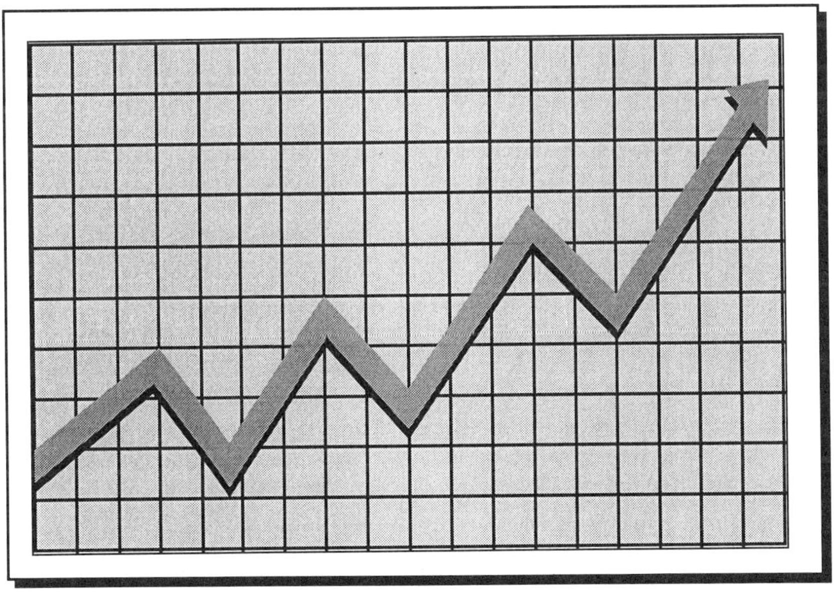

Continuous Improvement

to bring about innovation, including automation, shifts in technology, and state-of-the-art production processes. These mechanisms are based on the concept that an organization can be at one level of performance one day, and vastly improved the next. (Σi).

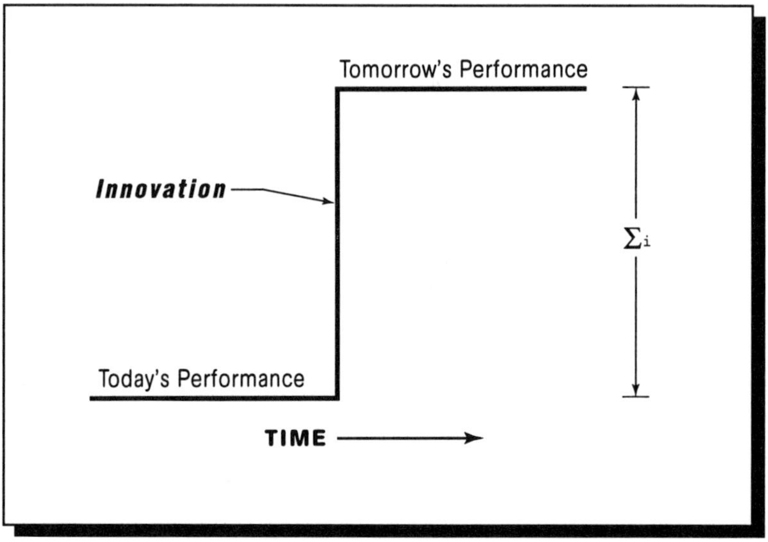

Figure 8.1. Quantum Leap Change Concept.

THE CONCEPT OF CONTINUOUS IMPROVEMENT

For our understanding of the continuous improvement step in change, it is important that we dissect the traditional American concept of improvement and offer an expanded definition.

Change is an ongoing part of today's and tomorrow's organizational environment. Consider that change is a step-by-step process for improvement. Each new change raises the efficiency and effectiveness of an organization or program. If you were to look closely at the construction of an ancient pyramid, you would see that each row of blocks forms a step-by-step assent toward the top. This is step-by-step improvement and is part of the organization's or program's continuous improvement strategy.

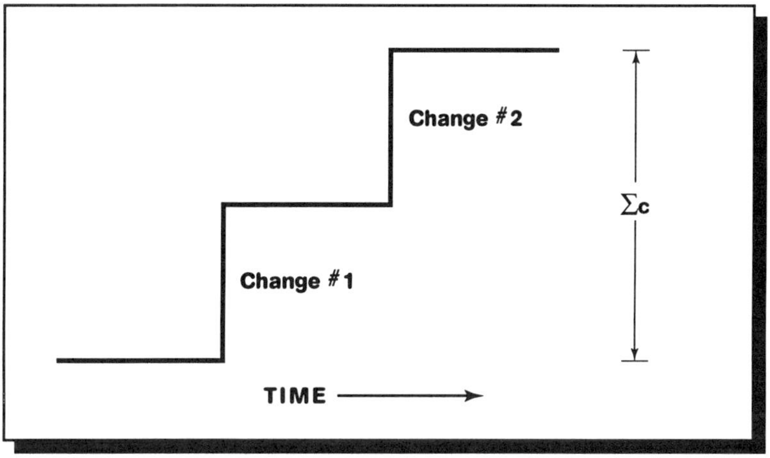

Figure 8.2. Step-by-Step Improvement Model.

But to understand the total picture of continuous improvement we must both back up to gain a larger, nondetailed view of our pyramid and look closer at each step or change. A view from farther off shows only a slope of improvement. The individual change efforts or steps are lost at this level of detail. This is one important aspect of the traditional model of continuous improvement.

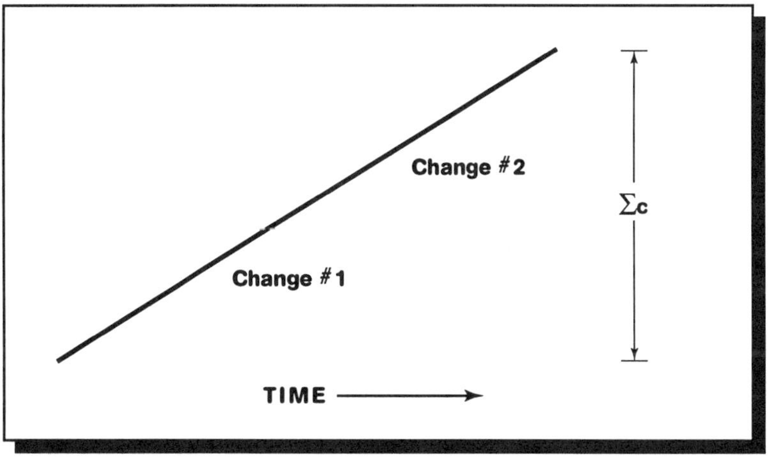

Figure 8.3. Traditional Model of Continuous Improvement.

In a process-oriented approach to continual improvement such as in total quality, we must also look at the new image of each step in this ongoing change environment. In this new environment, each change effort takes on individual continuous improvement. Unlike the horizontal step of the ancient pyramids, a step change in continuous improvement continues to rise following successful implementation. Thus, continuous improvement is not only advanced by each successive change implementation, the overall improvement is continually enhanced by the additive effects of each change's improvement (**Σci**).

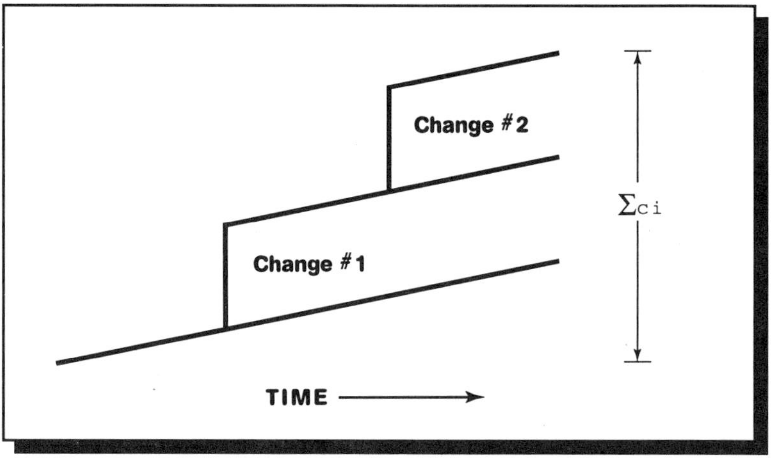

Figure 8.4. Successive Change Improvement Model.

A very different look at continuous improvement comes from a macro- and microperspective. There is an important differentiation to make between traditional step-by-step, resultsoriented improvement and process-oriented approaches. Because each change effort is in itself focused at continuous improvement, the total picture of change is altered in a process-oriented approach. Consequently, the resulting amplitude of improvement (**Σci**) is increased.

ANOTHER SHIFT IN THINKING

In America, we are absorbed by the desire to win. Our sports activities always have a winner. Our sales competitions always have a winner. Our contests always have a winner. We do not join the game merely to be participants. We all want to win the next Publisher's Clearinghouse™ giveaway or be the next one to win a million in Las Vegas. Our organizations, like us, are also focused on winning. As a slogan on a tee-shirt put it, "Second place is the first loser."

Lee Iacocca, then CEO of Chrysler, put the concept of winning a different way. He said, "To be the best, what else is there?" Notice the distinction between winning and being the best. Winning is fixed in time with a starting and finishing time. The purpose is to be ahead of the competition when the clock runs out. Being best, however, is focused at one point in time along a continuum. There is no reference to a starting or an ending time.

Why is this differentiation between winning and being best important in our discussion of continuous improvement? Because today's world is not measured in a designated time frame. It is a continuum of time.

Let me clarify with an example. Traditionally, we look at everything as if it were a race. It has a starting point in time when the gun is fired. It has an end when the winner first crosses the finish line. There is *one* winner, with a winning time. It's a singular, permanent concept. Being best is like a race that doesn't have a finish line. The measure of success is judged at each point in time by where you are compared to the pack of runners. If you are at the front, you are the best. But, being best at one moment guarantees you nothing in the next moment of the race. You can be passed by one or many other runners. Therefore, knowing the tenuous nature of being best, you must continually strive to stay in front and be the best at that point in time.

A confounder exists to the current race that makes the effort to be best even tougher. As the endless race proceeds, the competition improves. This creates an escalating improvement curve. All things remaining the same, an improvement curve would look like that shown as Figure 8.5 below. But, that's not the way today's environment is or tomorrow's will be. Everyone in the race will continue to improve or drop out, or new, improved competitors will enter. The resulting improvement curve will look more like Figure 8.6. As the competition improves, it increases the pressure on the continuous improvement stage to generate synergistic amplitude (**Σci'**).

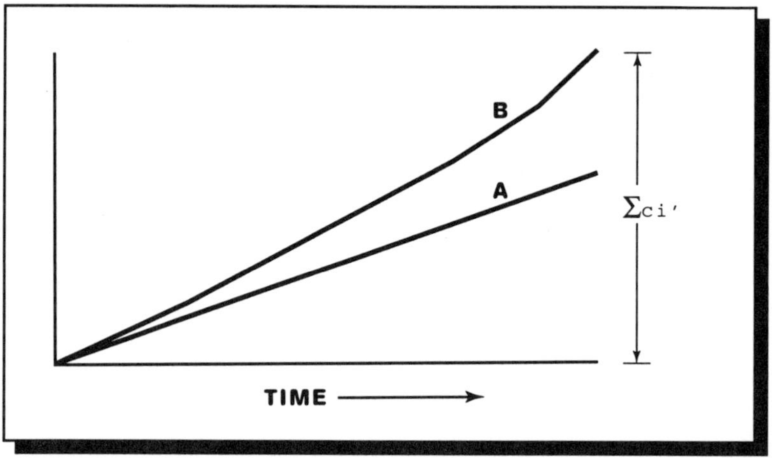

Figure 8.5. Competitive Escalating Improvement Model.

This is the complex environment of continuous improvement today. As you can see, it is very different from our traditional, accepted concept of rapidly winning through innovation. Today, the continuous improvement stage of change must focus on three aspects: 1) the macro-effort; step-by-step, change effort to next change effort, 2) the micro-effort; continued improvement of each implemented change, and 3) the competitive-effort of the ever-improving level of competition. The resulting graph of continuous improvement would look like Figure 8.6.

THE FOLLOW-THROUGH

The continuous improvement stage of change can be analogized to a portion of the process for hitting a tennis ball. The preparatory still is the mental aspect of deciding how you want to hit the ball, and the eye

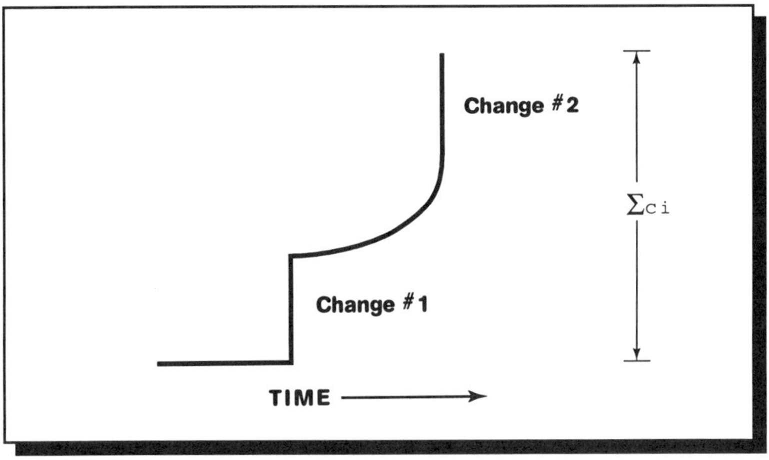

Figure 8.6. The Continuous Improvement Stage of Change.

contact with the oncoming fuzzy orb would be solicitating and creating an environment for change. The planning step is from the point you begin to bring your arm and racket back into a cocked position. Implementation is the swing toward the speeding ball to the point of impact. Impact is the actual accomplishment. The continuous improvement stage can be thought of as the follow-through as the racket moves through the ball, propelling it in the intended direction (you hope). As most tennis players know, the follow-through after you connect with and change the travel of the ball is one of the most important parts of a successful tennis swing. This is also true in the case of the continuous improvement stage in the change process.

REEVALUATING YOUR PROGRAM

Once a change has been successfully implemented, your program is altered, for the better. This shift becomes the new baseline for continuous improvement. Therefore, reevaluating your program is an important first step in this process and has two purposes—documenting improvement and identifying improvement opportunities.

Generally, performance measures used to monitor the change process document only implementation of the particular change. This is usually a specialized area of performance or increased activity. It can be very different from improvement in an area of your program. For example, in the conceptual stage of your change effort, you identified that you wanted to change the way your safety and health program operates from a traditional turf-basis to total quality. The performance measures you chose to monitor your change effort may have made use of a Gantt Chart, showing the various efforts and activities that needed to be implemented, in what order, and when.

You may have chosen to measure planning activities as an indicator of one of the key elements of total quality, and you may have measured line participation in safety and health activities and training. These are all good measures of a successful change to total quality. Once the change is implemented, going back and measuring the different aspects of your safety and health program would be very different.

You might follow the OSHA Voluntary Protection Program guidelines for evaluating your program, so you would rate it in areas such as employee involvement, hazard recognition and correction, formal procedures and policies, and staffing. This reevaluation following change provides a cross-reference to your implementation's true effectiveness. Did it really improve your program?

Successfully implementing a change effort that you sold the upper management is an important thing. But documenting that vision or change actually improved your program is critical. It tells upper management that you really knew what you were doing. It also tells them that future change efforts have a high probability of equal or better success because of your competence and leadership.

Documentation steps are critical in a business world where the pot is *always* being stirred by upper management. It encourages them to look at other programs than yours at cutback time.

The second purpose of reevaluating your program is to identify improvement opportunities. Just because a change (singular) has been successfully implemented, in today's environment of continual change, you need to constantly be looking how and where you can improve.

EVALUATION TOOLS

There are two approaches you can take in evaluating your safety and health program. Classified in a general sense, one uses outside sources and the other inside or expert sources. One is perception-driven. The other is more factual. The first is being evaluated by customers (external) and the second is using a formal program evaluation process (internal). See Table 8.1.

Table 8.1. Evaluation of a Safety Program.

Element	Customer Evaluation	Program Evaluation
Construction	Formal/written	Formal / written
Measurement	Perception	More factual
Focus	General program areas	Specific program elements
Information	Specific needs based on perception	Specific actions based on measurement

Both methods are formal and written. This is important because a large part of the power of evaluations comes from trends—the same question is asked time after time to see if performance is going up or down. Formal, therefore, means that the evaluation is carefully developed so that extensive changes are not needed from period to period. The rule is, if you change or add a question, your trend line starts again at the origin. You begin developing a trend after three evaluation periods. Evaluations are long-term processes.

The basis for measurement is different in external and internal evaluations. When you ask customers (those to whom you provide your functions or service) what they think about what you are doing, you ask for their perception. Like the customer feedback cards often found on restaurant tables or at fast food counters, the perception of the customer is important. Are you meeting the customers' needs for your service? Is the service timely or not?

Timing, of course, is a possible evaluation bias. If you ask at the time when things are going well, versus when they aren't, you have a much higher probability that the perception will be favorable. Keep in mind that evaluations seek "truth;" they are not tools to document only the positive side. If you want to use them for improvement, sometimes you have to face the negative perception. For this reason, customer evaluations are best accomplished on a given time schedule, mostly annually or with each service. Sticking with a regular time format gives you the best possible evaluation power and the least chance of bias.

Formal program evaluations provide a much more factual measurement because they are performed by those inside the program or those very knowledgeable about the specifics of the function. For this reason, the measurement will be based less on perceptive and more on fact.

The focus of each evaluation tool is also different. You just can't get program-specific in a nonexpert-based evaluation. Customers do not have the knowledge or skill to adequately evaluate the inner workings. Program evaluations done internally by experts yield very specific information because they have that knowledge and skill. For this reason, more specific questions can be asked and answered in formal program evaluations, such as, "Rate the knowledge and skill level of the safety and health program administration;" "Rate the effectiveness of the Safety Committee;" "Rate the commitment of management to safety and health;" "Rate the use of safety participation tools in the workplace;" and "Rate worker participation in safety."

The measurement criteria and scale become important. Where expert evaluation is used, measurement criteria need to be formally determined. What constitutes an excellent, high, or other rating? For example, if a Safety Committee meets at least monthly, publishes an agenda prior to each meeting, publishes minutes of each meeting within one week of the meeting, has a completion percentage of at least 80% of the year's agenda items, and has at least 90% attendance at the meetings, it should be rated as excellent. Additionally, there is a tremendous urge to add too many points or categories in the measurement scale. In fact, too many categories make rating much more difficult and the probability of mismeasurement (and incorrect trends) much higher. Try to not use more than five points or categories. Using numerical rating levels such as 5 to 1 or 4 to 1 is fine. Using ratings such as excellent, good, fair, and poor is also acceptable.

There are very different areas of information that you can get from the two approaches. In many ways, this becomes the reason for choosing one type over the other, or both. Because customer evaluations come from perception and nonexpert orientations, the information tends to be more needs-based. What are we doing that you don't like? What can we do better? If you aren't seeing us enough, how much would be just right? Whether items such as "get out into the areas more" are added to your action list for the upcoming year becomes a function of how many customers said they wanted more visibility of the safety group or the importance of the customer who said so. Because the view is perception-based, it shouldn't necessarily be taken as fact.

To a lesser extent, you can get action-specific information from customer evaluations. For example, you might ask, "what areas of safety training can we provide next year?"

Formal internal program evaluations almost always recommend specific actions based on the measurement. For example, the trend of the Safety Committee excellence measurement has gone from excellent last year to good this year. What are we not doing and what do we need to do in the upcoming year to get it back into the excellent area? This, in turn, becomes an action item on next year's program planning.

IMPROVING THE PROCESS

Once a change has been implemented, it can always be made better. This is *the* driving belief behind continuous improvement. We look at functions such as safety and health, as "programs." Traditionally, a program implies something that it is written, structured, not changing, and deeply entrenched. The continuous improvement stage examines all aspects of an organization (service, production, maintenance, etc.) as processes. A process is a dynamic concept. A process has a defined beginning and an ending point. It requires resources and takes time to accomplish.

This is the necessary perspective for us and our safety and health programs. A safety and health program is a process by which a necessary, value-adding function is provided for an organization. Like a manufacturing process that makes automobiles (macro-process), for example, it is comprised of many smaller processes (micro-processes). In this example, micro-processes include painting, transmission assembly, design, etc. Safety and health programs (macro-processes) also have micro-processes, including chemical approvals, accident investigations, exposure monitoring, ventilation evaluations, training and orientation, inspections, etc.

After a macro-process undergoes a successful change implementation such as to total quality, its basic structure is modified. The process is performed differently following the change. This new macro-process is the result of the macro-change effort. This new state or process becomes the starting place for continuous improvement. In continuous improvement, you first identify all of the micro-processes. Second, you identify the key micro-processes, which are those that take the most time and expertise to accomplish, are more frequently done, receive the most complaints, or cause the most frustration in your customers. This is a short list, perhaps two or three micro-processes. Third, you map each key micro-process by breaking it down into its steps or identifiable tasks and the sequence each step takes.

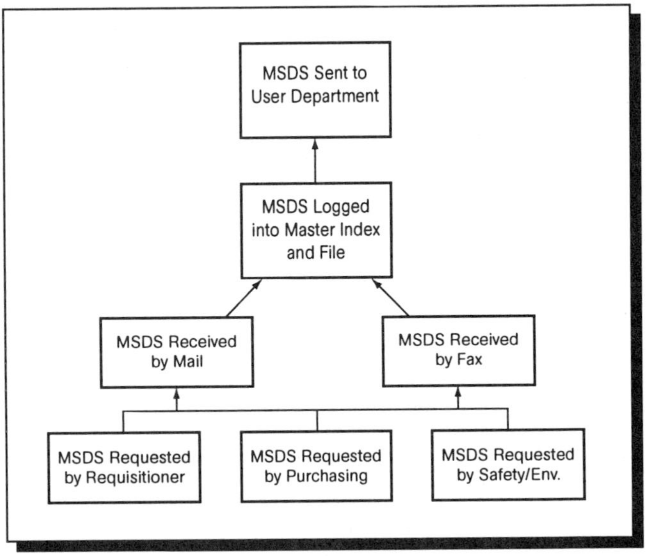

Figure 8.7. Vertical Task Breakdown.

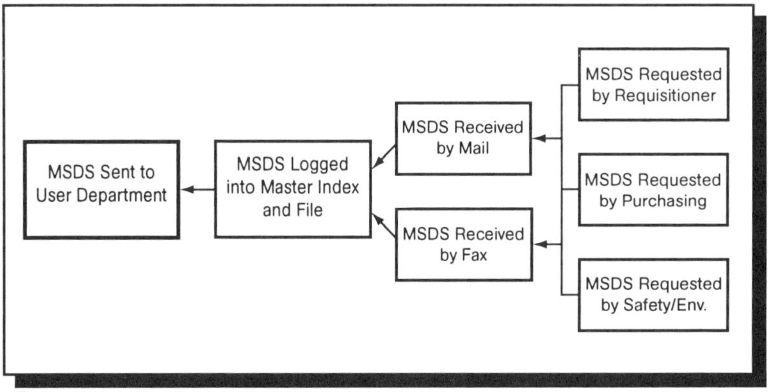

Figure 8.8. Process Mapping Tool.

Different methods that can be used to break down a micro-process. For example, a vertical task classification, a modified version of a fault-tree classification using "and/or" gates, or a sequential process mapping tool can be used.

Fourth, each task must be reviewed and the specifics of each task quantified. How long does that task take to accomplish (total task time)? How much actual hands-on time is devoted to each task, compared to the total amount of time the task takes (hands-on time)? Is there redundancy between any task and another (duplication)? Is the task necessary (value)? Can the task be done in a simpler or quicker way

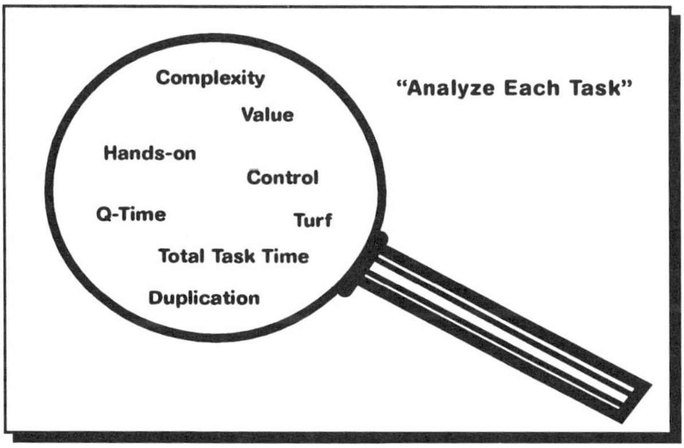

Figure 8.9. Analysis of Each Task.

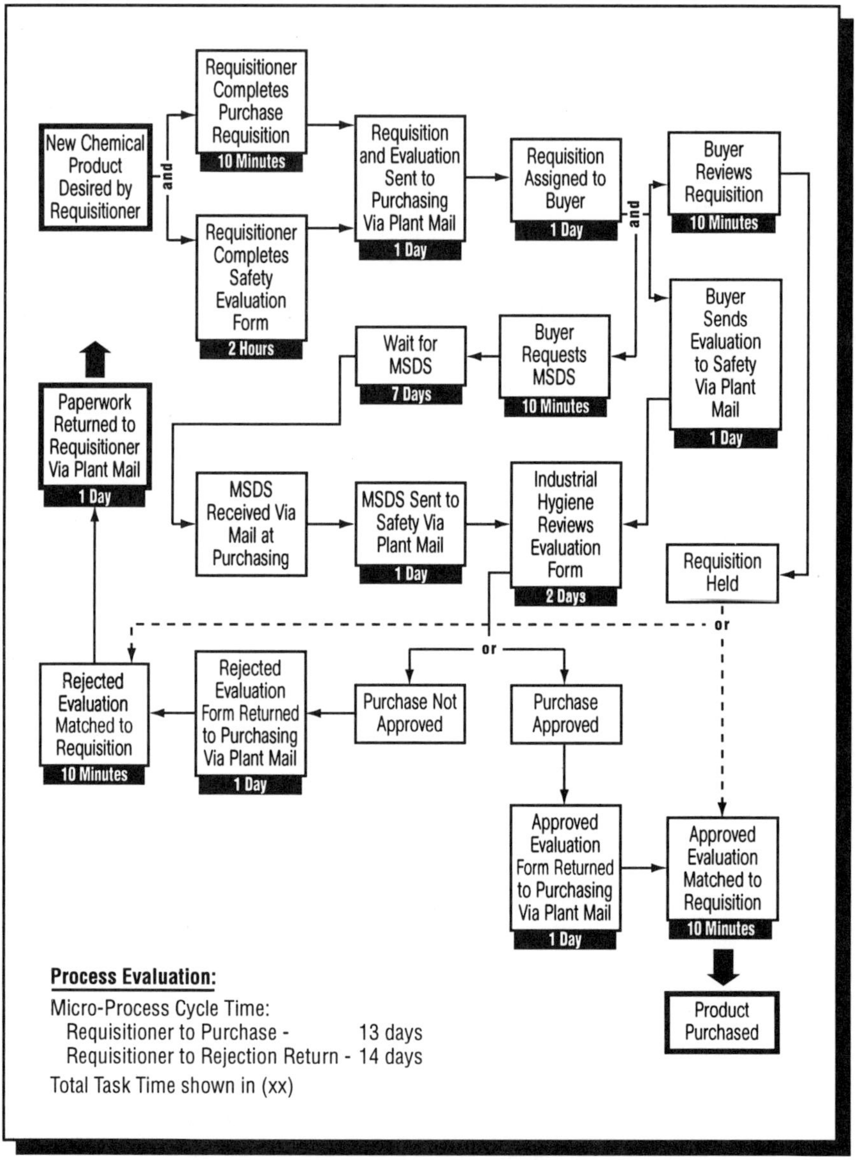

Figure 8.10. Example of Process Analysis Using Process Mapping Tool.

(complexity)? Is the task a bottleneck in which work stacks up waiting for one person, a particular specialty, an approval, or some other delay (Q-time)? Is there any reason that everyone can't do any one of the tasks (turf)? What training, skills, or control programs could you use to have almost anyone be able to do any task (control)?

The purpose of this continuous improvement stage is to improve the process. The analysis of the key micro-processes details where the micro-process can be improved. In the provided example of a process analysis, several areas of improvement can be easily identified using the task questioning approach. The use of electronic means to transmit the MSDS could cut the customer service time in half. The passage of materials between Purchasing and Safety can be streamlined to cut the resulting customer service time in half again. If industrial hygiene reviews evaluation forms at Purchasing, "on the fly," the customer service time can be cut by another two days. If a chemical evaluation form was required for those chemical products that meet published criteria, the need for a requisitioner to fill out the form would decrease considerably. This would increase the "customer friendliness" of the micro-process. If these simple changes are employed, a marked improvement can be realized. The total cycle time for the micro-process can be cut from 13–14 days to a total of 2 days, while improving the "customer friendliness" of the approval process.

Steps for the Continual Improvement

1. List the Micro-Processes
2. Identify the Key Micro-Processes
3. Break down Each Key Micro-Process into Tasks and Sequence
4. Analyze Each Task for Improvement Opportunities

Goals of Continuous Improvement

- Eliminate Steps that Don't Add Value
- Simplify the Process
- Increase the Efficiency of the Process
- Increase the Effectiveness of the Process

USING BENCHMARKING FOR CONTINUOUS IMPROVEMENT

We tend to look only inside of our program or organization for answers, when ample solutions are available outside. This is where benchmarking can be an effective tool for continuous improvement.

Once a key micro-process is identified, you can find someone else who is performing the task better and borrow the method. Because organizations and programs differ so greatly, however, benchmarking can represent a major change rather than a minor change, such as the kind of changes emphasized under continuous improvement. For this reason, benchmarking needs to be carefully done. Do not try to adopt the entire program—use the bite-sized approach. Look at specific micro-process tasks that can lead to improvement, and import ideas from outside your organization or program.

Used in this manner, benchmarking can have a significant effect since you import only what works. You can also import changes to many tasks and other micro-processes at the same time. This can decrease the amount of time necessary for significant improvement, and reduce the amount of time spent deliberating and choosing the right idea.

PARTNERSHIPPING OPPORTUNITIES

Partnershipping can be the source of ideas or of shared resources for continuous improvement. Partnershipping is no more than creating informal (and sometimes formal) information-sharing alliances. The most widely used vehicles for partnershipping are industry or trade organizations or associations, local like-industry gatherings can also be extremely effective. If you are looking for continuous improvement ideas, chances are that others are also. And with safety being a "one for all" concept, sharing your improvement ideas can help others who don't even recognize specific areas with potential for improvement.

A partnershipping effort within the OSHA Voluntary Protection Program (VPP) can serve dual purposes. The evaluation and application process can provide a valuable tool for reevaluating your safety and health program on a regular basis (every three years). It can also open doors, not only with OSHA, but with other industries in your area or state with which you can share ideas or solicit improvement strategies.

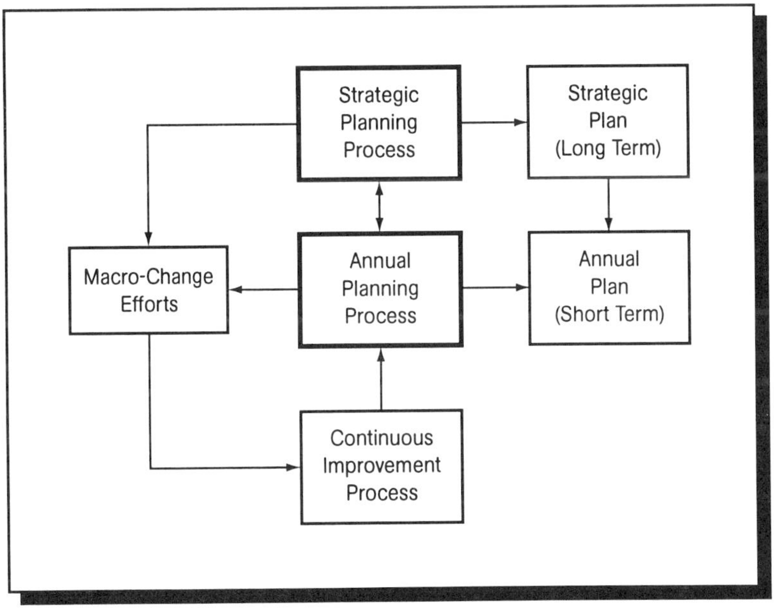

Figure 8.11. Total Quality Planning Process Including Macro-Change Effort.

THE DYNAMIC PLANNING PROCESS

Any improvement projects or actions must become part of your safety and health planning process. This is where programs that use total quality have a definite advantage. Planning is one of the basic driving concepts of total quality. This fits effectively with the continuous improvement stage of change.

Major change efforts come out of the strategic and annual planning processes. Once implementation is complete, the continuous improvement stage focuses on micro-change efforts. These improvement strategies, in turn, feed the planning process and are captured in the annual plan.

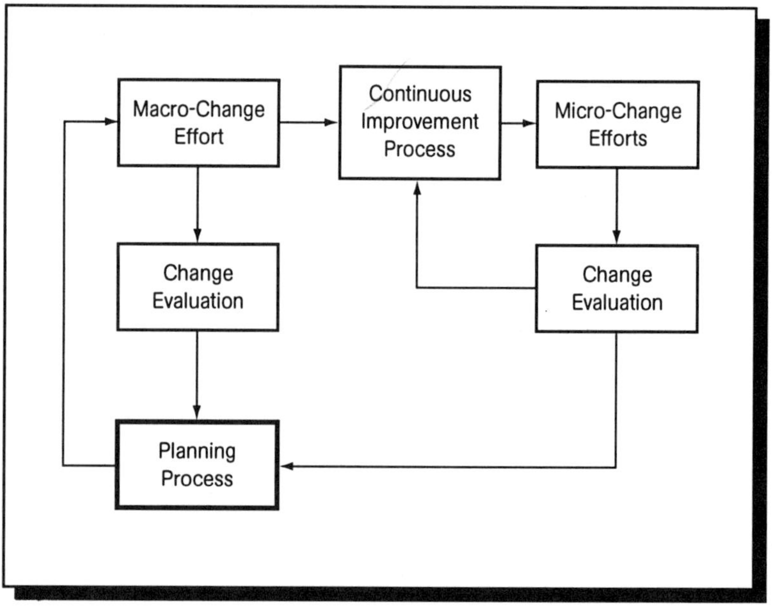

Figure 8.12. The Continuous Cycles of Change.

THE CYCLE OF CHANGE IN CONTINUOUS IMPROVEMENT

Continuous improvement is an ongoing process that fuels itself and can be displayed as a cycle within the cycle of change.

In the continuous improvement stage, we find the natural merge to total quality. In parallel with total quality concepts, the evaluation of change efforts, both macro and micro, can be provided by program evaluations and most importantly, measurement. As macro-change implementations follow through into continuous improvement, the planning elements of total quality guide the ongoing evolution of the effort. But all aspects of change are cyclical. Just as continuous improvement circles through change, evaluation, and further change, it feeds the planning cycle of major change. In two independent cycles, the small, no-cost or low-cost improvements of continuous improvement, and the major change efforts, exist in dependent harmony. Each feeds off of and contributes to the other in a synergistic relationship.

SUMMARY

An often misunderstood concept in American business, continuous improvement can provide many rewards to a change implementation effort. Change efforts provide step-by-step improvement. When change implementation is followed with continuous improvement, a linear ascension to each change effort is added. When each successful effort is combined with the ever-shifting baseline of improvement in competition, improvement becomes a near-linear upward slope. The overall process improvement (**ci**) is much greater than that provided by traditional innovation approaches (**i**) and macro-change alone (**c**).

The continuous improvement stage of change implementation also provides the ideal transition to total quality as a valuable management system approach for safety and health programs. Locked in dependent yet independent cycles, macro- and micro-change efforts provide an ongoing, synergistic approach to process improvement. With the use of analytical tools and a simple four-step approach, processes are broken down into tasks that, once visible, can be improved. This process-oriented approach to change has significant advantages over our traditional results-oriented approach via innovation. Where innovation is normally expensive, slow, and known to have a high failurc rate, process-oriented change is cheaper, quicker, and always yields improvement. Large, measurable improvements can be slow using continuous improvement alone. This has always been used as the major reason for rejecting continuous improvement as a viable method in America. But, when combined with macro-change implementation, continuous improvement provides a faster and more assured ascension.

9

A Checklist Approach to Successful Change

"Change? You're going to do what? When? How? Why???" Feel the blood pressure climb? The thought of skippering a major change effort through dark and rough seas, along reef-cluttered shores, into potential hurricane forces can bring the most fearless among us to near-panic. The reaction is natural and predictable, but manageable with the proper tools.

Change itself is a frightening but unavoidable reality today. Successfully implementing change is not simple. It is a very complex effort and requires great patience, preparation, and a dynamic planning approach. For these reasons, it is easy to fail somewhere along the implementation path. We've talked about some of the more common pitfalls including lack of adequate preparation. Ignoring details, being surprised by roadblocks and resistance, and not acquiring or keeping sponsors are also common dead ends to change efforts. Starting too soon or too late are additional stumbling points. But successfully implementing change is really as simple as following a step-by-step approach. Moving through each step in the change process, the important issues become visible and planning becomes the dynamic map to success.

In an effort to provide a map for successfully implementing change, the following checklist was developed. Used as a guide for a change implementation team or champion, it can serve as a sequential reminder of the important steps. Presented in a checklist format, it also clarifies the important steps that could be missed or hidden in the explanatory text of the previous chapters. The checklist, therefore, is both a clarifying and simplifying tool for change leaders.

Checklist for Change Implementation

Pre-Change Questions

Reference: Pre-change Question Flow Diagram

❑ Proceed with Change ❑ Action Needed

Action:

Stage I: Evaluation and Preparation

What is our history?	Action:
What is our culture?	Action:
What is our base line and our processes?	Action:
Why does my function exist: My Mission Statement:	Action:
Where do I want to go?	Action:

Checklist for Change Implementation — Page 2

Stage I: Evaluation and Preparation (Cont.)

When do we want to get there?	Action:
Who can we learn from?	Action:
Who are the critical players? The Change Implementation Team:	Action:
What skills do we need?	Action:
What will the impact be inside our program?	Action:
What will the impact be outside our program?	Action:

Checklist for Change Implementation — Page 3

Stage I: Evaluation and Preparation (Cont.)

What paradigms and culture stand in our way?	Action:
What resistance can we expect?	Action:
Is the path we've chosen correct? ❑ Yes ❑ Needs to be modified	Action:
What modifications are needed to ensure success?	Action:

Notes on Stage I: ______________________________

____

____

____

____

Checklist for Change Implementation — Page 4

Stage II: Solicitation and Creation of an Environment for Change

List appropriate Change Sponsors:	Specific Actions:
Sponsor A:	
Sponsor B:	
Sponsor C:	
Sponsor D:	

List appropriate Change Champions:	Specific Actions:
Champion A:	
Champion B:	
Champion C:	
Champion D:	

What resources will we need?

Training / Who?
Communication?
People?
Power?

Checklist for Change Implementation — Page 5

Stage II: Solicitation and Creation of an Environment for Change (Cont.)

Plan for entrenching our mission in the organization:

Plan	Who	When

Plan for sharing change vision with Sponsors:

Plan	Who	When

Plan for sharing change vision with the organization:

Plan	Who	When

What critical impacts of the change are necessary to share:

With Sponsors	Who	When

With Organization	Who	When

Checklist for Change Implementation — Page 6

Stage II: Solicitation and Creation of an Environment for Change (Cont.)

Plan for communicating critical impacts:

Plan	Who	When

List appropriate Change Agents:

Action:

Development/communication plan for Change Agents:

Plan	Who	When

What feedback pathways will we use?

Action:

Is modification of the plan necessary at this point?

☐ Yes ☐ No

If yes, what changes?

Action:

Checklist for Change Implementation — Page 7

Stage III: Planning and Implementation

What participation methods will we use in planning?	Action:

Plan for Implementation	Actual Implementation
What training is needed and when?	Scheduled / Completed:
What communication is needed and how?	Scheduled / Completed:
What measurement will be used?	Scheduled / Completed:
What review and approvals will be needed?	Scheduled / Completed:

Notes:

Checklist for Change Implementation — Page 8

Stage III: Planning and Implementation (Cont.)

Change Implementation Plan / Schedule:	Completed / Modifications:
Task and Sequence — Who	

Notes

Checklist for Change Implementation — Page 9

Stage III: Planning and Implementation (Cont.)

Change Implementation Plan / Schedule (Cont.):	Completed / Modifications:
Task and Sequence — Who	

Notes

Checklist for Change Implementation — Page 10

Stage IV: Transition

What is the plan for dealing with transition?	Actions / Notes on use:
Ongoing communication needs/pathways:	
Communicating/dealing with Sponsor needs:	
Identifying and dealing with resistance:	
Helping Change Agents:	
Restarting sluggish change progress:	

Checklist for Change Implementation — Page 11

Stage V: Celebration

How will we know when we've been successful?

What preparatory actions are necessary to make management easy about celebrating?

Who

Celebration Plan

Action

Notes

Checklist for Change Implementation — Page 12

Stage VI: Continuous Improvement

What method(s) of program evaluation will we use?

☐ Customer Evaluation

Detail plan and specifics: | How Implemented | When

☐ Formal Program Evaluation

What format will be used? | How Implemented | When

What process improvement tools will be used?

☐ Process Mapping

Process(es) to be mapped: | By Whom | When

☐ Other Methods:

Describe Method

Process(es) to be mapped: | By Whom | When

Checklist for Change Implementation — Page 13

Stage VI: Continuous Improvement (Cont.)

What measurements will we use to gauge progress?

Process	Meaurement(s)	Actions:

Will we use benchmarking or partnershipping?

☐ Yes ☐ No

If yes, detail:	Actions:

Notes

____

____

____

____

____

____

____

Checklist for Change Implementation —— Page 14

Post-Change: Change Evaluation / Lessons Learned

What parts of the change process/planning went well and which ones didn't?

Went Well	Struggled/Ineffective

Notes

Checklist for Change Implementation — Page 15

Post-Change: Change Evaluation / Lessons Learned (Cont.)

What lessons were learned to aid in the next change effort?

Notes

10

CHANGE—THE WAY OF THE FUTURE

Charles Darwin probably said it best: "It is not the strongest of the species that survive, nor the most intelligent, but the one most responsive to change." This quote is particularly relevant as we find ourselves in a world of continual and escalating change. A world in which we seem to be acted on rather than influencing the actions. But, we are not amoebas. We have the gift of choice. We are not simply subjected to change inside the evolutionary test tube of life. Our response to change is not as much biological in origin as mental. To a great extent, we choose our response to change.

Like the evolution mechanism that's been working for millions of years, there is a natural selection process today, but it's a lot faster. In evolution, organisms respond to changes in the environment. If the organism responds correctly, it survives; an incorrect response leads to death. This is evolutionary natural selection. Today, however, the need to change forces us to choose a response. Those with greater power determine whether or not our response was correct. If the response was correct, we are allowed a place in the changed organization. If, however, our response was not correct, we are set adrift inside or outside the organization. This is "organizational" natural selection.

A saying was popularized during a 1980s television ad campaign by the Chrysler Corporation. Then-Chrysler CEO, Lee Iacocca, stated, "In today's world, you can lead, follow, or get out of the way." In this concise statement lies the three roles we choose among when being confronted by change. How do we choose? Traditionally, we don't. We have a long history of choosing a fourth option—we ignore change. If we choose to lead, we are placed in a position where we cannot influence any of the impact or direction. If we choose not to follow, we potentially place ourselves in front of the change movement or direction. If we don't get out of the way, we are unprepared to deal with the force exerted by the change effort.

This was okay when management felt that we were necessary "costs of doing business." Our specialty justified our continued employment. But, in today's and especially in tomorrow's organizational environment, acceptance is not a given. With the increasing popularity of outsourcing specialized services, the traditional choice in change is detrimental to our future.

This book has been about change. It has mixed a vision of change from two different but interrelated aspects. The book provides a how-to oriented approach to mastering change efforts from a leadership or change champion vantage. But this book has also looked at our traditional stance on change.

In the 1930s, a Czechoslovakian-born economist, Joseph A. Schumpeter, described an economic process called "creative destruction." He deduced that customer demands cause improvement. Products are either made better or bigger, or more features are added. Services are improved, changed to be more customer friendly. Organizations aimed at improvement react to the demand of customers so they can better compete for business. Some changes are successful and some are not. Those who change successfully are rewarded by increased business, and those who do not change, struggle. When changes establish new customer expectations or new markets, organizations trapped with the old products and services don't survive. The customer simply no longer wants the product they make or the service they provide.

This is the economic phenomena that Schumpeter described. Some industries decline and die as others are being developed. The more aggressive (often younger) businesses replace the weaker or more entrenched ones. Those established firms that cannot adapt, or change unsuccessfully, are replaced by more visionary firms that satisfy customers. The economy recreates itself as old jobs give way to new ones. This is creative destruction.

What lesson does this theory teach us? Using today's real-world macro-economics lesson and applying it to our micro-economics situation, the lesson becomes clear. American business is caught in the reality of this global theater in which plays a world-wide version of creative destruction. As our organizations face competition, they must change to recapture customers and survive. The first battle a struggling organization faces is to control costs. This becomes an economic necessity. In this sphere, creative destruction has a micro-economics application. Specialized internal functions that serve the organization itself but do not directly bring in revenue to that organization are also locked in competition. In an environment where the costs of doing business must decrease, so-called nonvalue-adding functions come under critical evaluation. The heavier the economic burden, the more intense the critical evaluation. Just as some businesses will not survive due to macro-creative-destruction, some specialized functions won't survive due to micro-creative destruction.

So, what is the best choice we can make concerning our involvement in change? We need to be leading the charge. Leadership and involvement are the keys to successful employment opportunities in tomorrow's organizational world. Those who choose to step out of the way or to ignore change will suffer the ultimate insult of creative destruction. This is why having a tool or approach for mastering change is important to us. We need skills or tools that can make us successful, especially if we choose to lead the change effort. In this role, there is no place to hide. But, because we can be successful leading change, we can create our own job security.

POINTS TO REMEMBER

Mastering change is a product of preparation, planning, and dynamic implementation. We've discussed a six-stage approach to successful change implementation. The approach begins with a questioning sequence for assessing environment and organizational roadblocks that would derail change before it gets on the track. This questioning sequence ends in a "go" or "no go" decision before extensive preparation or planning efforts are expended. The six stages of the change process include evaluation and preparation, solicitation and creation of an environment for change, planning and implementation, transition celebration, and continuous improvement. Emphasized in the change implementation checklist is an evaluation of the completed effort and collection of any wisdom gained. This lessons-learned exercise provides valuable input into improving the actual change process. You might look at it as a continuous improvement tool that you would find in a total quality program.

This is the process for mastering change. But, there are some critical points that successful change masters must never forget. Let's discuss eight.

Eight Points You Can Never Forget!

1. You Never "Arrive"
2. Communication Is Never Wasted
3. Training Must Precede Expectation
4. Participation Is Critical
5. Synergy Is THE Power
6. Letting Go Is the Key
7. There Is a Constant Threat of Resurgence of Traditionalism
8. American Business Is Cyclic

You Never "Arrive." In today's ever-spiraling world of change and improvement, *you never "arrive."* Remember our discussions about how continuous improvement and competition change the rules of today's game? Today's rules do not address winning. Today's rules only address staying ahead of the pack. There is a constant demand for improvement and continual effort. It is not determined by the market, only by your position in the competitive pack. Your position right now guarantees you nothing in the next second.

I liken the current business and inner organizational environment to the old tale of Jack and the Beanstalk, only the rules are different. The new rules say the giant is on the ground and climbing up after you.

They also say that the beanstalk never stops growing. To stay ahead of the probable doom (should the giant catch-up to you), you must continually climb. As you climb, you adapt and the constant workout increases your muscle mass. This helps you climb faster (a form of continual improvement). But, just when you think that you may be able to pull away from the giant and get some deserved rest, he too gets stronger and climbs faster. The advantage you've gained melts away.

Communication Is Never Wasted. Communication is a strange thing. It's like guessing how many and what kinds of clothes to take on a trip. If you pack just right, you always have the appropriate things to wear while away. If, however, you forget to pack something, or you pack clothes for 70° and get 45°, or you end up needing a swimsuit or maybe a tuxedo, you are woefully unprepared. To counter this, wise travelers take too many clothes. You may never use some of them, but you never lack the desired or appropriate apparel.

Another strange thing about communication is that the proper amount is never determined by the transmitter—the person who knows the information and needs to share it. It is only determined by the receiver—the person who does not know and who needs the information. When we are in the role of the transmitter, we forget this. We fall victim to traditional communication paradigms and outdated communication rules. "They can't understand that;" "You can't tell them everything;" "They don't have a need to know;" or "It must be officially blessed before communicated." The "information is power" rule is a controlling, stratifying element in communication. These are all roadblocks to effective communication.

The trick to effectively communicating in change is to give *too much* information. You want to give them so much information that it answers *any* question they might ever have. You want them to be able to throw away some of the information. Because, if you don't give them enough, the mind tends to create data in the blank spaces. Usually, this created information is negative. It comes from the "if they didn't tell me, it is probably bad" idea. Negative information also acts as a magnet to the grape vine or rumor mill. Once negative information is transferred via this pathway, it is impossible to manage, correct, or turn positive. You can only prevent the creation of negative information by looking at communication from the needs of the receiver and discarding traditional communication limiting paradigms.

Training Must Precede Expectation. We have a strange, disjointed, but widely traditional belief in America: "Anyone can do that." It's one of our ultimate assumptions. "If I want them to change a light bulb, anyone can do that," "If I want them to work as a team, anyone can do that," or "If I want them to read and comprehend a set of work rules, anyone can do that,"

The problem is that over and over again we discover that everyone *can't* do that. *Training must precede expectation.* When we assume that knowledge, ability, and skill exist, we set ourselves up for failure. This is especially true in today's increasingly complex world.

If we want people to work effectively in teams, we must invest in giving them team skills. If we want them to take on challenging responsibilities such as being a Change Agent, we must teach them how to be successful. If change means that people are going to have to do new things or do old things using new ways, we must train them to do so.

Training cannot fall victim to assumption, being overlooked, or stubborn resistance. If you don't assure that those involved in change or impacted by it are trained so that they can be successful, your change effort will fail because they will fail.

Participation Is Critical. There are some simple yet strong correlations that link participation with successful change implementation.

CORRELATION #1: The more people that work toward change, the greater the probability the change will be successfully implemented.

More People = Higher Success Rate

CORRELATION #2: The more people you have working toward change, the faster the change can be successfully implemented.

More People = Less Time

As you would probably expect, opposing correlations also exist. If you try to "drive" change through by yourself, you are in for a lot of frustration, a low probability of success, and the effort (if you can sustain your own motivation) will take forever.

Participation in change, therefore, isn't just important, it's critical to the change's success. If you have no forces, you cannot lead. In participation, the greater the numbers, the greater your opportunity to lead. It is through the combined actions of the masses that significant change comes about. As examples, consider the changes in the former Soviet Union, the destructive changes in the area of the world formerly called Yugoslavia, the civil rights movement in this country, and the democratic changes in South Africa. None of these happened because one person, or even a small group of people, came together and forced the change through. It takes the masses to bring about large-scale change, positive or negative. Participation can be your biggest strength in the success of your change effort or the lack of it can be the anchor that brings the change progress to a screeching halt.

Synergy Is THE Power. Synergy is when the combined result of something is greater than the summation of all its parts. A formula displaying synergistic results would be nonsensical to a mathematician (1 + 1+ 1 + 1 + 1 = 8). This is synergy. In a change process where the participation level is high, you can build synergy by using empowering tools such as making sure everyone has the knowledge and skills to be successful via training and information. Communication is another empowering tool. Through communication you can better use, motivate, and coordinate available forces for change. Measurement also builds synergy through motivation. Those involved in the change effort are motivated by progress. The final empowering tool is involving all players in a team effort. There exists tremendous power and support forces in team structures. It is through the team concept that you can gain the greatest synergy for your change effort.

Letting Go Is the Key. This is a tough one for those who accept leadership roles. Traditionally, we think of a leader as someone who takes charge of a situation, a challenge, or effort. Is leadership at the head of the charging force? No—leadership is a "behind the scenes" effort. There are times, especially at the start of a change effort, when the change champion is at the lead selling the change vision and talking about the benefits. But, throughout the change effort, the champion works to direct and support the team effort. For example, during the D-Day invasion of Europe during World War II, Eisenhower lead his forces by being on the first landing craft at Normandy. Was he the first to jump behind enemy lines in support of the invasion? Absolutely not. During the invasion, he was in England planning, communicating, coordinating, and motivating. This is a leadership role.

True leaders inspire and enable teams to work toward visions, and then get out of the way. A leader who insists on being at the head of the pack all of the time and taking the credit can choke the energy out of a change effort. Soon enough, that type of leader will be the only one working toward change. This makes him or her a zealot in the eyes of the organization.

There Is a Constant Threat of Resurgence of Traditionalism. Keep in mind our discussions about the transition stage of change and the predictable emotional response cycle to change. These constant threats are a product of looking backward. The purpose of transition is to change the orientation of those impacted by change from the past to the future. Recognizing that a major change effort has to force those impacted by the change into the lowest emotional levels, depression, before their minds begin to explore the future possibilities, transition can be very tricky and demanding. It is this natural past-orientation that is the root of the threat of resurgence of traditionalism.

A scene from the epic movie, *The Ten Commandments,* is especially useful to emphasizing this threat. The Israelites have just left bondage in Egypt. They were treated as subhumans, beaten, continually worked past exhaustion, and killed. It was a terrible existence. But, now they are free for the first time. Being stranded in the desert, being backed-up against the Red Sea, and seeing Pharaoh's army approaching, the first reaction was to long for what was known: bondage in Egypt. It mattered little that it was awful. It also mattered little that they would have to cast aside an age-old vision of being led out of bondage—the very promise that had given them hope for years and years. In this depressed and threatened state, they could only look back.

This is just where transition finds change efforts. A change leader must never forget this constant threat that can undermine the change effort so quickly. It is especially a concern when you recognize that, as a change leader, you must drive them through depression to gain a future-oriented outlook.

There are three significant sources of this focus on the past: 1) management thought and practice, 2) labor isolation and power, and 3) our safety and health turf.

Management thought and practice. To those of us who have followed and worked in American business for most of our careers, the term "forward-thinking management" seems to be an oxymoron. Our history and our business management reaction during the 1980s and 1990s, when America's industry suffered serious defeats from the expanding global economy, stands as testimony that a much closer definition is exactly the opposite. In fact, "head-in-the-sand," and "past-oriented" are standard phrases for describing American management. Traditionalism, good or bad, is revered as sound business practice.

This is a tremendous threat in itself because change is sponsored by management. Having sponsorship constantly at odds with tradition places large burdens on change efforts. Past-orientation is most prevalent in transition, when emotional response is at its lowest and the needs for firm commitment and continued resources are at their highest.

Labor isolation and power. Two interrelated sources in labor contribute to the potential for providing a threat of resurgence of traditionalism. The first is labor's traditional organizational isolation.

This is not just an American division. Many believe that you either think like management or you think like labor.

Labor is both an oppressed and an oppressing force. Regardless of what organization-wide change takes place, labor is generally most impacted by it. In the power equation, labor is generally a sleeping giant. When organized, however, its power can be impressive. But, don't get caught in the traditional way of thinking about organized labor. Labor can be organized by something like a union or employee association. These are formal organizations of labor. Labor can also be organized by common emotion. This is informal organization and generally only lasts as long as the emotional response does. For example, give everybody a 10% pay cut and watch them organize emotionally.

Change begets emotion. If enough people perceive a negative impact, you will get a collective emotional response. This response can, and often does, translate to significant power in the labor force—those impacted by the change. Whether you focus change at a work team or at an entire organization, the extent of that power is directly proportionate to the percentage impacted with negative emotion. For example, if a change impacts 2,000 workers and approximately two-thirds have a negative perception of that change, you can expect meeting with considerable resistance (about two-thirds of the total is possible) in foot-dragging, change-stopping, labor power. Likewise, if you were impacting a work group of six and four were collectively very negative about the change. Your potential negative power is two-thirds, the same as when the 2000 were impacted. In any case, the change would meet tremendous resistance and probably stop in its tracks.

How does this potential for labor power feed the threat of resurgence of traditionalism? As in the thought, practice, and emotional response of management, labor has potentially a high level of power, not from sponsorship roots, but from shear numbers. Forcing targets into a low emotional level to gain future-oriented thinking must first traverse the domain of past-oriented thinking. Low collective emotional response and the past-orientation combine to form a significant threat of resurgence to the way things were.

Our safety and health turf. We need anchors in life and work upon which to hang our needs for security. We have to have a certain diet of predictability to be well adjusted. When everything is changing in someone's life, physical and emotional problems are common. Traditionally, we have used our turf as a means of predictability. Turf may include our office or cubical, our staff (if any), our job description and assigned responsibilities, or tools such as reference materials and industrial hygiene equipment.

Change means that things are done differently. If change impacts our turf, we get caught in the emotion trap where we can long for what was. This problem occurs less often, however, when we have championed the change effort. But, we must be honest in our assessment of the impacts of change to our turf. This allows us to deal with it before it is tangible, and to adjust to it. If we are not completely honest and downplay the impact on our turf, we can have a magnified emotional response that is brought about by shock. If we, as change champions, suffer emotionally, the change process will. Resurgence of traditionalism and return of the predictable turf becomes the capitulation.

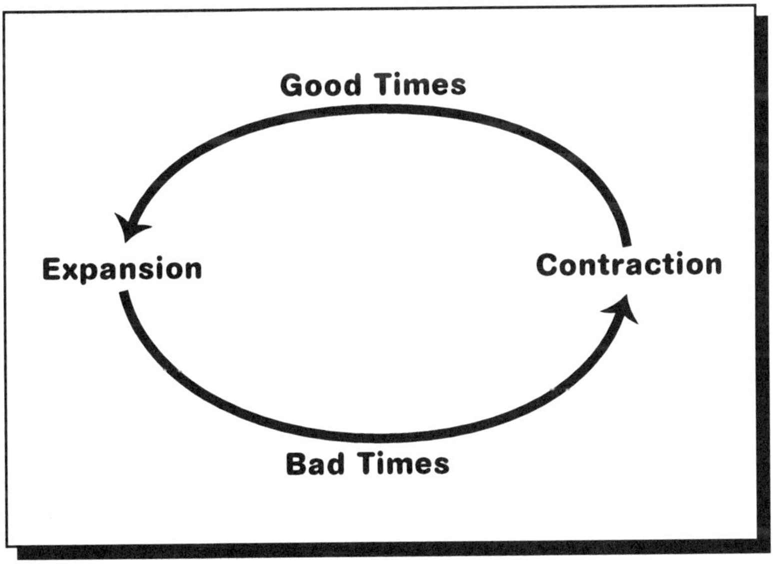

Figure 10.1. The Traditional American Organizational Cycle.

American Business Is Cyclic. From an organizational focus, American business seems to be locked in an inescapable cycle of organizational ups and downs. The organization expands during prosperity (both business and economic) and shrinks during downturns.

Expansion is slow; contraction is rapid. Expansion tends to occur across the entire organizational structure—workers, management, and staff. Contraction has historically focused mainly on workers and staff; workers, because the downturn made less production or service necessary, and staff, because they are considered nonvalue-adding areas of the organization.

The latest round of business downsizing has been a little different. As is traditional, it has focused on workers and staff areas of the organization. But this time it has streamlined middle management. The standard organizational pyramid has experienced a lot of compression as organizations have sought to adopt the flat organizational model. Another significant change concerning this current right-sizing effort in American business is the extent of staff streamlining. In prior cycles, the number of staff positions in safety and health has decreased. Once good times reappear, this has allowed the predictable cycle to again expand those staff numbers. This time, however, the right-sizing efforts are questioning staff functions at a much deeper level. Instead of questioning the number of staff needed to provide a specialized function, current emphasis places the very survival of the function within the organization in question. Upper management is questioning whether it is better economically if the specialty is contracted when and if it is needed. With cuts of this magnitude, the function has to face significantly more challenges should the expansion period of the organizational cycle occur again. Having no function in the organizational structure means no voice when expansion.

This new depth of organizational streamlining can have a significant impact if a particular staff or specialty function is contracted out. Because the function is now outside of the organization and outside of the decision-making sphere, the impact can be detrimental. Therefore, in future business decision-making, the out-sourced specialty only receives visibility as cash, a debit on the ledger. This plays directly into the traditional economic-focus, the Drucker model, of management. The question then becomes "Can we afford this service or not?" The need for that specialty takes a distant rear seat in this type of questioning process. In most cases, having no in-house champion, the question of need isn't even asked.

The Downsizing Hit List (Listed in Traditional Order)

1. Workers
2. Maintenance Trades
3. Staff Tied to Perks or Benefits (Laundry, Cafeteria, Fitness/Health Related, Etc.)
4. Value-Disconnected Staff and Functions (Product Assurance, Safety/Health, HR, Engineering, Etc.)
5. First-Line Management
6. Essential Staff Personnel
7. Management
8. Upper Management

When the function is only streamlined in a downsizing effort, the voice on the inside is a standing testimony to the need or perceived need by the organization. This foothold translates to upward adjustments in the size of that function when economic upswings, good times, and expansion occur. Safety and health personnel have traditionally been at the head of the downsizing list. Under this deeper level of downsizing, if the safety and health function is out-sourced, it has a much greater chance of never regaining the ground that was lost. The safety and health function then becomes an economics-only issue to the organization. Consequently, the mission for improving worker safety and health suffers.

TOTAL QUALITY AS *THE* PRACTICE FOR THE FUTURE

In business downturns, total quality can be both a blessing and a curse. It is definitely a blessing in that it works so well and I know of no other management method that works better in safety and health programs. Ideally, total quality seats the responsibility of worker safety and health where it belongs and where it can be successful—in the line structure. It marries responsibility with true authority. Total quality can also provide invaluable tools for helping a safety and health program reach excellence and document value to the organization. These are real advantages to the safety and health program, the level of worker safety and health in the organization, and to the economics and efficiency of that organization.

But, this is a mixed signal. We need to begin this discussion with an important clarifying point. Staff functions do not make any money directly for the organization. Realistically, they are only debits on the ledger. They can add significant value to that organization, especially if the staff function uses an effective management concept like total quality, and that brings about a seating of the responsibility for worker safety and health in the line structure. But, the real purpose of *any* non-profit-contributing function is to deeply seat what they do in the profit-contributing portion of the organization to such an extent that the need for the specialized staffing disappears. Realistically, like a truly flat organizational structure, I question if this ideal can ever be reached. But in a true waste elimination, continuous improvement strategy it must be the ultimate goal.

The curse with total quality can come in a downsizing effort in which management perceives that this ideal has been met, or when the economic realities are so frightening and require such drastic cuts that management decides to take the risk. This is not totally negative, however. Experience tells us that the innovative leaders who adopted new, more effective management methods like total quality are retained in the organization. They are usually given new opportunities and they are often the last to leave. This can be a significant positive aspect because of both continued employment and the presence of an "inner voice" to the organization should expansion occur.

In any event, downsizing doesn't seem to adversely impact the level of worker safety and health in the organization. This remains constant because it is a function of the seating of worker safety and health in the line structure *and* the continuance of the safety and health function somewhere else in the organization.

BEYOND TOTAL QUALITY

Total quality is only the foundation. It is a highly effective management method. But integral to total quality is the concept of continuous improvement. And in continuous improvement is the constant search for new tools, advanced concepts, and the management models of the future. What are they? Where will they be discovered? Who will take them from rough concept to successful practice?

As with many new ways of thinking, new management methods and concepts are coming from outside normal channels. But one doesn't have to look to the future. Similar concepts have been proven successful in other arenas. They just are not well known to American management. When total quality burst onto the American management scene in the mid-1980s, it had been practiced and refined for 20 years in Japanese businesses.

Another Japanese management concept that is receiving a lot of attention in American businesses is time-based management, or TBM. TBM considers time to be the most precious and nonreplenishable of resources. Analytical tools such as process mapping are used to identify the steps of each process. Each process is broken down into the total time it takes to complete. Time when actual work is done is called hands-on time. Time spent waiting is Q-time. The purpose of TBM is to eliminate tasks that are not necessary, remove the Q-time, and reduce the hands-on time. Accomplishing this, the entire cycle time is shortened and the other necessary resources are minimized.

Is TBM applicable to safety and health programs? Of course it is! It is a further expansion of the process mapping tool that was discussed in the continuous improvement stage of change.

Japanese management literature describes many techniques and concepts that have had little introduction in American organizations.

Many concepts may offer great advantages to safety and health program management. Additionally, new concepts such as chaos theory and creative morphology offer ideas that are yet to be explored.

IMPORTANCE OF OUR ROLE IN CHANGE

Change is a reality today and tomorrow. It will continue to invade our world at home, at work, and at play with increasing frequency and range. Because of it, what is considered unthinkable today will be common place tomorrow. This is exciting.

At no time in our past has the threat to the practice of safety and health been greater. In our changing world where nothing is left unquestioned, our very existence and our purpose are under constant attack. It isn't just a war out there, it is a fight for survival. But, with the stakes so high, we can no longer afford to wait for the storm to blow over. We stand too high a risk of failing in this approach. The choice is truly ours to make. Will we choose to ignore the obvious changes all around us, or turn and run toward the light? Will we lead positive change and contribute to our successful destiny? Will we be known in our organization and to our decision-makers as progressive leaders? Will we seek out new ways of thinking and pioneer the changes from rough concept to successful reality?

The advantages are many. The risks are high. But, in tomorrow's world, the risks of not changing will be even higher. The question is, given a successful approach to successfully implementing change, what will we do with our future?

References

Barker, J.A. *Paradigm Pioneers*. Burnsville, MN: Chart House International, 1994.

Barker, J.A. *The Business of Paradigms: Discovering the Future*. Burnsville, MN: Chart House International, 1990.

Connor, D.R. *Managing at the Speed of Change*. New York: Villard, 1992.

Coonradt, C.A. *The Game of Work*. Salt Lake City, UT: Shadow Mountain, 1984.

Hammer, M. and Champy, J. *Reengineering the Corporation: A Manifesto for Business Revolution*. New York: Harper, 1993.

Imai, M. *Kaizen: The Key to Japan's Competitive Success*. New York: Random House, 1986.

Kubler-Ross, E. *On Death and Dying*. New York: Macmillan, 1969.

Larkin, T.J. and Larkin, S. *Communicating Change: Winning Employee Support for New Business Goals*. New York: McGraw-Hill, 1994.

Ohno, T. *Workplace Mangement*. Cambridge, MA: Productivity Press, 1982.

Pierce, F. D. *Total Quality for Safety and Health Professionals*. Rockville, MD: Government Institutes, Inc., 1995.

Pierce, F. D. *Shifting Safety and Health Paradigms*. Rockville, MD: Government Institutes, Inc., 1996

Pritchett, P. *New Work Habits for a Radically Changing World*. Dallas, TX: Pritchett & Associates, 1994.

Shiba, S. Graham, A., and Walden, D. *A New American TQM: Four Practical Revolutions in Management*. Portland, OR: Productivity Press, 1993.

Singo, S. *The Sayings of Shigeo Shingo: Key Strategies for Plant Improvement.* Portland, OR: Productivity Press, 1985.

Slater, R. *Get Better or Get Beaten: 31 Leadership Secrets from GE's Jack Welch.* Burr Ridge, IL: Irwin, 1994.

Smith, P.G.,and Reinertsen, D.G. *Developing Products in Half the Time.* New York: Van Nostrand Reinhold, 1991.

Stalk Jr., G. and Hout, T.M. *Competing Against Time: How Time-Based Competition is Reshaping Global Markets.* New York: Free Press, 1990.

Thompson Jr., L. *Mastering the Challenges of Change: Strategies for Each Stage in Your Organization's Life Cycle.* New York: American Management Association, 1994.

Wheatley, M. J. *Leadership and the New Science.* San Francisco: Berrett-Koehler Publishers, 1992.

INDEX

D

E

M

N

O

P

T

U

V

A

X